Geocomputation with R

Chapman & Hall/CRC
The R Series

Series Editors

John M. Chambers, Department of Statistics, Stanford University, California, USA

Torsten Hothorn, Division of Biostatistics, University of Zurich, Switzerland

Duncan Temple Lang, Department of Statistics, University of California, Davis, USA

Hadley Wickham, RStudio, Boston, Massachusetts, USA

For more information about this series, please visit: https://www.crcpress.com/go/the-r-series

Geocomputation with R

Robin Lovelace
Jakub Nowosad
Jannes Muenchow

CRC Press
Taylor & Francis Group
Boca Raton London New York

CRC Press is an imprint of the
Taylor & Francis Group, an **informa** business
A CHAPMAN & HALL BOOK

CRC Press
Taylor & Francis Group
6000 Broken Sound Parkway NW, Suite 300
Boca Raton, FL 33487-2742

First issued in paperback 2020

© 2019 by Taylor & Francis Group, LLC
CRC Press is an imprint of Taylor & Francis Group, an Informa business

No claim to original U.S. Government works
Version Date: 20190220

ISBN 13: 978-0-367-67057-3 (pbk)
ISBN 13: 978-1-138-30451-2 (hbk)

Visit the Taylor & Francis Web site at
http://www.taylorandfrancis.com

and the CRC Press Web site at
http://www.crcpress.com

For Katy

Dla Jagody

Für meine Katharina und alle unsere Kinder

Contents

Foreword

Doing 'spatial' in R has always been about being broad, seeking to provide and integrate tools from geography, geoinformatics, geocomputation and spatial statistics for anyone interested in joining in: joining in asking interesting questions, contributing fruitful research questions, and writing and improving code. That is, doing 'spatial' in R has always included open source code, open data and reproducibility.

Doing 'spatial' in R has also sought to be open to interaction with many branches of applied spatial data analysis, and also to implement new advances in data representation and methods of analysis to expose them to cross-disciplinary scrutiny. As this book demonstrates, there are often alternative workflows from similar data to similar results, and we may learn from comparisons with how others create and understand their workflows. This includes learning from similar communities around Open Source GIS and complementary languages such as Python, Java and so on.

R's wide range of spatial capabilities would never have evolved without people willing to share what they were creating or adapting. This might include teaching materials, software, research practices (reproducible research, open data), and combinations of these. R users have also benefitted greatly from 'upstream' open source geo libraries such as GDAL, GEOS and PROJ.

This book is a clear example that, if you are curious and willing to join in, you can find things that need doing and that match your aptitudes. With advances in data representation and workflow alternatives, and ever increasing numbers of new users often without applied quantitative command-line exposure, a book of this kind has really been needed. Despite the effort involved, the authors have supported each other in pressing forward to publication.

So, this fresh book is ready to go; its authors have tried it out during many tutorials and workshops, so readers and instructors will be able to benefit from knowing that the contents have been and continue to be tried out on people like them. Engage with the authors and the wider R-spatial community, see value in having more choice in building your workflows and most important, enjoy applying what you learn here to things you care about.

Roger Bivand

Bergen, September 2018

Preface

This book is for people who want to analyze, visualize and model geographic data with open source software. It is based on R, a statistical programming language that has powerful data processing, visualization and geospatial capabilities. The book covers a wide range of topics and will be of interest to a wide range of people from many different backgrounds, especially:

- People who have learned spatial analysis skills using a desktop Geographic Information System (GIS) such as QGIS[1], ArcMap[2], GRASS[3] or SAGA[4], who want access to a powerful (geo)statistical and visualization programming language and the benefits of a command-line approach (Sherman, 2008):

> With the advent of 'modern' GIS software, most people want to point and click their way through life. That's good, but there is a tremendous amount of flexibility and power waiting for you with the command line.

- Graduate students and researchers from fields specializing in geographic data including Geography, Remote Sensing, Planning, GIS and Geographic Data Science
- Academics and post-graduate students working on projects in fields including Geology, Regional Science, Biology and Ecology, Agricultural Sciences (precision farming), Archaeology, Epidemiology, Transport Modeling, and broadly defined Data Science which require the power and flexibility of R for their research
- Applied researchers and analysts in public, private or third-sector organizations who need the reproducibility, speed and flexibility of a command-line language such as R in applications dealing with spatial data as diverse as

---

[1] http://qgis.org/en/site/
[2] http://desktop.arcgis.com/en/arcmap/
[3] https://grass.osgeo.org/
[4] http://www.saga-gis.org/en/index.html

Urban and Transport Planning, Logistics, Geo-marketing (store location analysis) and Emergency Planning

The book is designed for intermediate-to-advanced R users interested in geo-computation and R beginners who have prior experience with geographic data. If you are new to both R and geographic data, do not be discouraged: we provide links to further materials and describe the nature of spatial data from a beginner's perspective in Chapter 2 and in links provided below.

How to read this book

The book is divided into three parts:

1. Part I: Foundations, aimed at getting you up-to-speed with geographic data in R.
2. Part II: Extensions, which covers advanced techniques.
3. Part III: Applications, to real-world problems.

The chapters get progressively harder in each so we recommend reading the book in order. A major barrier to geographical analysis in R is its steep learning curve. The chapters in Part I aim to address this by providing reproducible code on simple datasets that should ease the process of getting started.

An important aspect of the book from a teaching/learning perspective is the **exercises** at the end of each chapter. Completing these will develop your skills and equip you with the confidence needed to tackle a range of geospatial problems. Solutions to the exercises, and a number of extended examples, are provided on the book's supporting website, at geocompr.github.io[5].

Impatient readers are welcome to dive straight into the practical examples, starting in Chapter 2. However, we recommend reading about the wider context of *Geocomputation with R* in Chapter 1 first. If you are new to R, we also recommend learning more about the language before attempting to run the code chunks provided in each chapter (unless you're reading the book for an understanding of the concepts). Fortunately for R beginners R has a supportive community that has developed a wealth of resources that can help. We particularly recommend three tutorials: R for Data Science[6] (Grolemund and Wickham, 2016) and Efficient R Programming[7] (Gillespie and Lovelace, 2016), especially Chapter 2[8] (on installing and setting-up R/RStudio) and

---

[5]https://geocompr.github.io/

[6]http://r4ds.had.co.nz/

[7]https://csgillespie.github.io/efficientR/

[8]https://csgillespie.github.io/efficientR/set-up.html#r-version

Chapter 10[9] (on learning to learn), and An introduction to R[10] (Venables et al., 2017). A good interactive tutorial is DataCamp's Introduction to R[11].

Why R?

Although R has a steep learning curve, the command-line approach advocated in this book can quickly pay off. As you'll learn in subsequent chapters, R is an effective tool for tackling a wide range of geographic data challenges. We expect that, with practice, R will become the program of choice in your geospatial toolbox for many applications. Typing and executing commands at the command-line is, in many cases, faster than pointing-and-clicking around the graphical user interface (GUI) a desktop GIS. For some applications such as Spatial Statistics and modeling R may be the *only* realistic way to get the work done.

As outlined in Section 1.2, there are many reasons for using R for geocomputation: R is well-suited to the interactive use required in many geographic data analysis workflows compared with other languages. R excels in the rapidly growing fields of Data Science (which includes data carpentry, statistical learning techniques and data visualization) and Big Data (via efficient interfaces to databases and distributed computing systems). Furthermore R enables a reproducible workflow: sharing scripts underlying your analysis will allow others to build-on your work. To ensure reproducibility in this book we have made its source code available at github.com/Robinlovelace/geocompr[12]. There you will find script files in the `code/` folder that generate figures: when code generating a figure is not provided in the main text of the book, the name of the script file that generated it is provided in the caption (see for example the caption for Figure 12.2).

Other languages such as Python, Java and C++ can be used for geocomputation and there are excellent resources for learning geocomputation *without R*, as discussed in Section 1.3. None of these provide the unique combination of package ecosystem, statistical capabilities, visualization options, powerful IDEs offered by the R community. Furthermore, by teaching how to use one language (R) in depth, this book will equip you with the concepts and confidence needed to do geocomputation in other languages.

Geocomputation with R will equip you with knowledge and skills to tackle a wide range of issues, including those with scientific, societal and environmental

---

[9] https://csgillespie.github.io/efficientR/learning.html
[10] http://colinfay.me/intro-to-r/
[11] https://www.datacamp.com/courses/free-introduction-to-r
[12] https://github.com/Robinlovelace/geocompr#geocomputation-with-r

implications, manifested in geographic data. As described in Section 1.1, geocomputation is not only about using computers to process geographic data: it is also about real-world impact. If you are interested in the wider context and motivations behind this book, read on; these are covered in Chapter 1.

Acknowledgements

Many thanks to everyone who contributed directly and indirectly via the code hosting and collaboration site GitHub, including the following people who contributed direct via pull requests: katygregg, erstearns, eyesofbambi, tyluRp, marcosci, mdsumner, rsbivand, pat-s, gisma, ateucher, annakrystalli, gavin-simpson, Himanshuteli, yutannihilation, katiejolly, layik, mvl22, nickbearman, ganes1410, richfitz, SymbolixAU, wdearden, yihui, chihinl, gregor-d, p-kono, pokyah. Special thanks to Marco Sciaini, who not only created the front cover image, but also published the code that generated it (see `frontcover.R` in the book's GitHub repo). Dozens more people contributed online, by raising and commenting on issues, and by providing feedback via social media. The `#geocompr` hashtag will live on!

We would like to thank John Kimmel from CRC Press, who has worked with us over two years to take our ideas from an early book plan into production via four rounds of peer review. The reviewers deserve special mention here: their detailed feedback and expertise substantially improved the book's structure and content.

We thank Patrick Schratz and Alexander Brenning from the University of Jena for fruitful discussions on and input into Chapters 11 and 14. We thank Emmanuel Blondel from the Food and Agriculture Organization of the United Nations for expert input into the section on web services; Michael Sumner for critical input into many areas of the book, especially the discussion of algorithms in Chapter 10; Tim Appelhans and David Cooley for key contributions to the visualization chapter (Chapter 8); and Katy Gregg, who proofread every chapter and greatly improved the readability of the book.

Countless others could be mentioned who contributed in myriad ways. The final thank you is for all the software developers who make geocomputation with R possible. Edzer Pebesma (who created the **sf** package), Robert Hijmans (who created **raster**) and Roger Bivand (who laid the foundations for much R-spatial software) have made high performance geographic computing possible in R.

1

Introduction

This book is about using the power of computers to *do things* with geographic data. It teaches a range of spatial skills, including: reading, writing and manipulating geographic data; making static and interactive maps; applying geocomputation to solve real-world problems; and modeling geographic phenomena. By demonstrating how various geographic operations can be linked, in reproducible 'code chunks' that intersperse the prose, the book also teaches a transparent and thus scientific workflow. Learning how to use the wealth of geospatial tools available from the R command line can be exciting, but creating *new ones* can be truly liberating. Using the command-line driven approach taught throughout, and programming techniques covered in Chapter 10, can help remove constraints on your creativity imposed by software. After reading the book and completing the exercises, you should therefore feel empowered with a strong understanding of the possibilities opened up by R's impressive geographic capabilities, new skills to solve real-world problems with geographic data, and the ability to communicate your work with maps and reproducible code.

Over the last few decades free and open source software for geospatial (FOSS4G) has progressed at an astonishing rate. Thanks to organizations such as OSGeo, geographic data analysis is no longer the preserve of those with expensive hardware and software: anyone can now download and run high-performance spatial libraries. Open source Geographic Information Systems (GIS) such as QGIS[1] have made geographic analysis accessible worldwide. GIS programs tend to emphasize graphical user interfaces (GUIs), with the unintended consequence of discouraging reproducibility (although many can be used from the command line as we'll see in Chapter 9). R, by contrast, emphasizes the command line interface (CLI). A simplistic comparison between the different approaches is illustrated in Table 1.1.

This book is motivated by the importance of reproducibility for scientific research (see the note below). It aims to make reproducible geographic data analysis workflows more accessible, and demonstrate the power of open geospatial software available from the command-line. "Interfaces to other software are part of R" (Eddelbuettel and Balamuta, 2018). This means that in addition to outstanding 'in house' capabilities, R allows access to many other spatial

---

[1]http://qgis.org/en/site/

TABLE 1.1: Differences in emphasis between software packages (Graphical User Interface (GUI) of Geographic Information Systems (GIS) and R).

Attribute	Desktop GIS (GUI)	R
Home disciplines	Geography	Computing, Statistics
Software focus	Graphical User Interface	Command line
Reproducibility	Minimal	Maximal

software libraries, explained in Section 1.2 and demonstrated in Chapter 9. Before going into the details of the software, however, it is worth taking a step back and thinking about what we mean by geocomputation.

Reproducibility is a major advantage of command-line interfaces, but what does it mean in practice? We define it as follows: "A process in which the same results can be generated by others using publicly accessible code."

This may sound simple and easy to achieve (which it is if you carefully maintain your R code in script files), but has profound implications for teaching and the scientific process (Pebesma et al., 2012).

1.1   What is geocomputation?

Geocomputation is a young term, dating back to the first conference on the subject in 1996.[2] What distinguished geocomputation from the (at the time) commonly used term 'quantitative geography', its early advocates proposed, was its emphasis on "creative and experimental" applications (Longley et al., 1998) and the development of new tools and methods (Openshaw and Abrahart, 2000): "GeoComputation is about using the various different types of geodata and about developing relevant geo-tools within the overall context of a 'scientific' approach." This book aims to go beyond teaching methods and code; by the end of it you should be able to use your geocomputational skills, to do "practical work that is beneficial or useful" (Openshaw and Abrahart, 2000).

Our approach differs from early adopters such as Stan Openshaw, however, in its emphasis on reproducibility and collaboration. At the turn of the 21st Century, it was unrealistic to expect readers to be able to reproduce code

---

[2]The conference took place at the University of Leeds, where one of the authors (Robin) is currently based. The 21st GeoComputation conference was also hosted at the University of Leeds, during which Robin and Jakub presented, led a workshop on 'tidy' spatial data analysis and collaborated on the book (see www.geocomputation.org for more on the conference series, and papers/presentations spanning two decades).

examples, due to barriers preventing access to the necessary hardware, software and data. Fast-forward two decades and things have progressed rapidly. Anyone with access to a laptop with ~4GB RAM can realistically expect to be able to install and run software for geocomputation on publicly accessible datasets, which are more widely available than ever before (as we will see in Chapter 7).[3] Unlike early works in the field, all the work presented in this book is reproducible using code and example data supplied alongside the book, in R packages such as **spData**, the installation of which is covered in Chapter 2.

Geocomputation is closely related to other terms including: Geographic Information Science (GIScience); Geomatics; Geoinformatics; Spatial Information Science; Geoinformation Engineering (Longley, 2015); and Geographic Data Science (GDS). Each term shares an emphasis on a 'scientific' (implying reproducible and falsifiable) approach influenced by GIS, although their origins and main fields of application differ. GDS, for example, emphasizes 'data science' skills and large datasets, while Geoinformatics tends to focus on data structures. But the overlaps between the terms are larger than the differences between them and we use geocomputation as a rough synonym encapsulating all of them: they all seek to use geographic data for applied scientific work. Unlike early users of the term, however, we do not seek to imply that there is any cohesive academic field called 'Geocomputation' (or 'GeoComputation' as Stan Openshaw called it). Instead, we define the term as follows: working with geographic data in a computational way, focusing on code, reproducibility and modularity.

Geocomputation is a recent term but is influenced by old ideas. It can be seen as a part of Geography, which has a 2000+ year history (Talbert, 2014); and an extension of *Geographic Information Systems* (GIS) (Neteler and Mitasova, 2008), which emerged in the 1960s (Coppock and Rhind, 1991).

Geography has played an important role in explaining and influencing humanity's relationship with the natural world long before the invention of the computer, however. Alexander von Humboldt's travels to South America in the early 1800s illustrates this role: not only did the resulting observations lay the foundations for the traditions of physical and plant geography, they also paved the way towards policies to protect the natural world (Wulf, 2015). This book aims to contribute to the 'Geographic Tradition' (Livingstone, 1992) by harnessing the power of modern computers and open source software.

The book's links to older disciplines were reflected in suggested titles for the book: *Geography with R* and *R for GIS*. Each has advantages. The former

---

[3]A laptop with 4GB running a modern operating system such as Ubuntu 16.04 onward should also be able to reproduce the contents of this book. A laptop with this specification or above can be acquired second-hand for ~US$100 in many countries nowadays, reducing the financial/hardware barrier to geocomputation far below the levels in operation in the early 2000s, when high-performance computers were unaffordable for most people.

conveys the message that it comprises much more than just spatial data: non-spatial attribute data are inevitably interwoven with geometry data, and Geography is about more than where something is on the map. The latter communicates that this is a book about using R as a GIS, to perform spatial operations on *geographic data* (Bivand et al., 2013). However, the term GIS conveys some connotations (see Table 1.1) which simply fail to communicate one of R's greatest strengths: its console-based ability to seamlessly switch between geographic and non-geographic data processing, modeling and visualization tasks. By contrast, the term geocomputation implies reproducible and creative programming. Of course, (geocomputational) algorithms are powerful tools that can become highly complex. However, all algorithms are composed of smaller parts. By teaching you its foundations and underlying structure, we aim to empower you to create your own innovative solutions to geographic data problems.

1.2   Why use R for geocomputation?

Early geographers used a variety of tools including barometers, compasses and sextants[4] to advance knowledge about the world (Wulf, 2015). It was only with the invention of the marine chronometer[5] in 1761 that it became possible to calculate longitude at sea, enabling ships to take more direct routes.

Nowadays such lack of geographic data is hard to imagine. Every smartphone has a global positioning (GPS) receiver and a multitude of sensors on devices ranging from satellites and semi-autonomous vehicles to citizen scientists incessantly measure every part of the world. The rate of data produced is overwhelming. An autonomous vehicle, for example, can generate 100 GB of data per day (The Economist, 2016). Remote sensing data from satellites has become too large to analyze the corresponding data with a single computer, leading to initiatives such as OpenEO[6].

This 'geodata revolution' drives demand for high performance computer hardware and efficient, scalable software to handle and extract signal from the noise, to understand and perhaps change the world. Spatial databases enable storage and generation of manageable subsets from the vast geographic data stores, making interfaces for gaining knowledge from them vital tools for the future. R is one such tool, with advanced analysis, modeling and visualization capabilities. In this context the focus of the book is not on the language itself (see Wickham, 2014a). Instead we use R as a 'tool for the trade' for

---

[4]https://en.wikipedia.org/wiki/Sextant
[5]https://en.wikipedia.org/wiki/Marine_chronometer
[6]http://r-spatial.org/2016/11/29/openeo.html

understanding the world, similar to Humboldt's use of tools to gain a deep understanding of nature in all its complexity and interconnections (see Wulf, 2015). Although programming can seem like a reductionist activity, the aim is to teach geocomputation with R not only for fun, but for understanding the world.

R is a multi-platform, open source language and environment for statistical computing and graphics (r-project.org/[7]). With a wide range of packages, R also supports advanced geospatial statistics, modeling and visualization. New integrated development environments (IDEs) such as RStudio have made R more user-friendly for many, easing map making with a panel dedicated to interactive visualization.

At its core, R is an object-oriented, functional programming language[8] (Wickham, 2014a), and was specifically designed as an interactive interface to other software (Chambers, 2016). The latter also includes many 'bridges' to a treasure trove of GIS software, 'geolibraries' and functions (see Chapter 9). It is thus ideal for quickly creating 'geo-tools', without needing to master lower level languages (compared to R) such as C, FORTRAN or Java (see Section 1.3). This can feel like breaking free from the metaphorical 'glass ceiling' imposed by GUI-based or proprietary geographic information systems (see Table 1.1 for a definition of GUI). Furthermore, R facilitates access to other languages: the packages **Rcpp** and **reticulate** enable access to C++ and Python code, for example. This means R can be used as a 'bridge' to a wide range of geospatial programs (see Section 1.3).

Another example showing R's flexibility and evolving geographic capabilities is interactive map making. As we'll see in Chapter 8, the statement that R has "limited interactive [plotting] facilities" (Bivand et al., 2013) is no longer true. This is demonstrated by the following code chunk, which creates Figure 1.1 (the functions that generate the plot are covered in Section 8.4).

```
library(leaflet)
popup = c("Robin", "Jakub", "Jannes")
leaflet() %>%
  addProviderTiles("NASAGIBS.ViirsEarthAtNight2012") %>%
  addMarkers(lng = c(-3, 23, 11),
             lat = c(52, 53, 49),
             popup = popup)
```

It would have been difficult to produce Figure 1.1 using R a few years ago, let alone as an interactive map. This illustrates R's flexibility and how, thanks to developments such as **knitr** and **leaflet**, it can be used as an interface to other software, a theme that will recur throughout this book. The use of

---

[7]https://www.r-project.org/
[8]http://adv-r.had.co.nz/Functional-programming.html

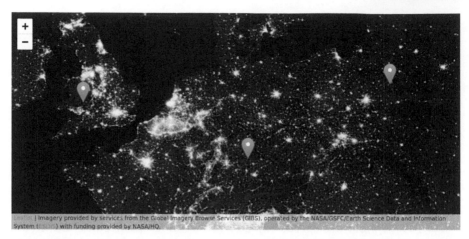

FIGURE 1.1: The blue markers indicate where the authors are from. The basemap is a tiled image of the Earth at night provided by NASA. Interact with the online version at geocompr.robinlovelace.net, for example by zooming in and clicking on the popups.

R code, therefore, enables teaching geocomputation with reference to reproducible examples such as that provided in Figure 1.1 rather than abstract concepts.

1.3 Software for geocomputation

R is a powerful language for geocomputation but there are many other options for geographic data analysis providing thousands of geographic functions. Awareness of other languages for geocomputation will help decide when a different tool may be more appropriate for a specific task, and place R in the wider geospatial ecosystem. This section briefly introduces the languages C++[9], Java[10] and Python[11] for geocomputation, in preparation for Chapter 9.

An important feature of R (and Python) is that it is an interpreted language. This is advantageous because it enables interactive programming in a Read–Eval–Print Loop (REPL): code entered into the console is immediately executed and the result is printed, rather than waiting for the intermediate

---

[9] https://isocpp.org/
[10] https://www.oracle.com/java/index.html
[11] https://www.python.org/

stage of compilation. On the other hand, compiled languages such as C++ and Java tend to run faster (once they have been compiled).

C++ provides the basis for many GIS packages such as QGIS[12], GRASS[13] and SAGA[14] so it is a sensible starting point. Well-written C++ is very fast, making it a good choice for performance-critical applications such as processing large geographic datasets, but is harder to learn than Python or R. C++ has become more accessible with the **Rcpp** package, which provides a good 'way in' to C programming for R users. Proficiency with such low-level languages opens the possibility of creating new, high-performance 'geoalgorithms' and a better understanding of how GIS software works (see Chapter 10).

Java is another important and versatile language for geocomputation. GIS packages gvSig, OpenJump and uDig are all written in Java. There are many GIS libraries written in Java, including GeoTools and JTS, the Java Topology Suite (GEOS is a C++ port of JTS). Furthermore, many map server applications use Java including Geoserver/Geonode, deegree and 52°North WPS.

Java's object-oriented syntax is similar to that of C++. A major advantage of Java is that it is platform-independent (which is unusual for a compiled language) and is highly scalable, making it a suitable language for IDEs such as RStudio, with which this book was written. Java has fewer tools for statistical modeling and visualization than Python or R, although it can be used for data science (Brzustowicz, 2017).

Python is an important language for geocomputation especially because many Desktop GIS such as GRASS, SAGA and QGIS provide a Python API (see Chapter 9). Like R, it is a popular[15] tool for data science. Both languages are object-oriented, and have many areas of overlap, leading to initiatives such as the **reticulate** package that facilitates access to Python from R and the Ursa Labs[16] initiative to support portable libraries to the benefit of the entire open source data science ecosystem.

In practice both R and Python have their strengths and to some extent which you use is less important than the domain of application and communication of results. Learning either will provide a head-start in learning the other. However, there are major advantages of R over Python for geocomputation. This includes its much better support of the geographic data models vector and raster in the language itself (see Chapter 2) and corresponding visualization possibilities (see Chapters 2 and 8). Equally important, R has unparalleled support for statistics, including spatial statistics, with hundreds of packages (unmatched by Python) supporting thousands of statistical methods.

---

[12] www.qgis.org
[13] https://grass.osgeo.org/
[14] www.saga-gis.org
[15] https://stackoverflow.blog/2017/10/10/impressive-growth-r/
[16] https://ursalabs.org/

The major advantage of Python is that it is a *general-purpose* programming language. It is used in many domains, including desktop software, computer games, websites and data science. Python is often the only shared language between different (geocomputation) communities and can be seen as the 'glue' that holds many GIS programs together. Many geoalgorithms, including those in QGIS and ArcMap, can be accessed from the Python command line, making it well-suited as a starter language for command-line GIS.[17]

For spatial statistics and predictive modeling, however, R is second-to-none. This does not mean you must choose either R or Python: Python supports most common statistical techniques (though R tends to support new developments in spatial statistics earlier) and many concepts learned from Python can be applied to the R world. Like R, Python also supports geographic data analysis and manipulation with packages such as **osgeo**, **Shapely**, **NumPy** and **PyGeoProcessing** (Garrard, 2016).

1.4   R's spatial ecosystem

There are many ways to handle geographic data in R, with dozens of packages in the area.[18] In this book we endeavor to teach the state-of-the-art in the field whilst ensuring that the methods are future-proof. Like many areas of software development, R's spatial ecosystem is rapidly evolving (Figure 1.2). Because R is open source, these developments can easily build on previous work, by 'standing on the shoulders of giants', as Isaac Newton put it in 1675[19]. This approach is advantageous because it encourages collaboration and avoids 'reinventing the wheel'. The package **sf** (covered in Chapter 2), for example, builds on its predecessor **sp**.

A surge in development time (and interest) in 'R-spatial' has followed the award of a grant by the R Consortium for the development of support for Simple Features, an open-source standard and model to store and access vector geometries. This resulted in the **sf** package (covered in Section 2.2.1). Multiple places reflect the immense interest in **sf**. This is especially true for the R-sig-Geo Archives[20], a long-standing open access email list containing much R-spatial wisdom accumulated over the years.

It is noteworthy that shifts in the wider R community, as exemplified by the data

---

[17]Python modules providing access to geoalgorithms include `grass.script` for GRASS, `saga-python` for SAGA-GIS, `processing` for QGIS and `arcpy` for ArcGIS.

[18]An overview of R's spatial ecosystem can be found in the CRAN Task View on the Analysis of Spatial Data (see `https://cran.r-project.org/web/views/Spatial.html`).

[19]`http://digitallibrary.hsp.org/index.php/Detail/Object/Show/object_id/9285`

[20]`https://stat.ethz.ch/pipermail/r-sig-geo/`

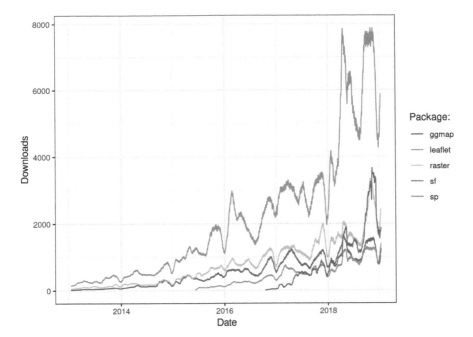

FIGURE 1.2: The popularity of spatial packages in R. The y-axis shows average number of downloads per day, within a 30-day rolling window, of prominent spatial packages.

processing package **dplyr** (released in 2014[21]) influenced shifts in R's spatial ecosystem. Alongside other packages that have a shared style and emphasis on 'tidy data' (including, e.g., **ggplot2**), **dplyr** was placed in the **tidyverse** 'metapackage' in late 2016[22]. The **tidyverse** approach, with its focus on long-form data and fast intuitively named functions, has become immensely popular. This has led to a demand for 'tidy geographic data' which has been partly met by **sf** and other approaches such as **tabularaster**. An obvious feature of the **tidyverse** is the tendency for packages to work in harmony. There is no equivalent **geoverse**, but there are attempts at harmonization between packages hosted in the r-spatial[23] organization and a growing number of packages use **sf** (Table 1.2).

---

[21] https://cran.r-project.org/src/contrib/Archive/dplyr/
[22] https://cran.r-project.org/src/contrib/Archive/tidyverse/
[23] https://github.com/r-spatial/discuss/issues/11

TABLE 1.2: The top 5 most downloaded packages that depend on sf, in terms of average number of downloads per day over the previous month. As of 2019-01-29 there are 117 packages which import sf.

package	Downloads
ggplot2	22593
plotly	3628
raster	2433
leaflet	1391
spData	1174

1.5   The history of R-spatial

There are many benefits of using recent spatial packages such as **sf**, but it also important to be aware of the history of R's spatial capabilities: many functions, use-cases and teaching material are contained in older packages. These can still be useful today, provided you know where to look.

R's spatial capabilities originated in early spatial packages in the S language (Bivand and Gebhardt, 2000). The 1990s saw the development of numerous S scripts and a handful of packages for spatial statistics. R packages arose from these and by 2000 there were R packages for various spatial methods "point pattern analysis, geostatistics, exploratory spatial data analysis and spatial econometrics", according to an article[24] presented at GeoComputation 2000 (Bivand and Neteler, 2000). Some of these, notably **spatial**, **sgeostat** and **splancs** are still available on CRAN (Rowlingson and Diggle, 1993, 2017; Venables and Ripley, 2002; Majure and Gebhardt, 2016).

A subsequent article in R News (the predecessor of The R Journal[25]) contained an overview of spatial statistical software in R at the time, much of which was based on previous code written for S/S-PLUS (Ripley, 2001). This overview described packages for spatial smoothing and interpolation, including **akima** and **geoR** (Akima and Gebhardt, 2016; Jr and Diggle, 2016), and point pattern analysis, including **splancs** (Rowlingson and Diggle, 2017) and **spatstat** (Baddeley et al., 2015).

The following R News issue (Volume 1/3) put spatial packages in the spotlight again, with a more detailed introduction to **splancs** and a commentary on future prospects regarding spatial statistics (Bivand, 2001). Additionally, the issue introduced two packages for testing spatial autocorrelation that eventually became part of **spdep** (Bivand, 2017). Notably, the commentary mentions

---

[24] http://www.geocomputation.org/2000/GC009/Gc009.htm
[25] https://journal.r-project.org/

the need for standardization of spatial interfaces, efficient mechanisms for exchanging data with GIS, and handling of spatial metadata such as coordinate reference systems (CRS).

maptools (written by Nicholas Lewin-Koh; Bivand and Lewin-Koh, 2017) is another important package from this time. Initially **maptools** just contained a wrapper around shapelib[26] and permitted the reading of ESRI Shapefiles into geometry nested lists. The corresponding and nowadays obsolete S3 class called "Map" stored this list alongside an attribute data frame. The work on the "Map" class representation was nevertheless important since it directly fed into **sp** prior to its publication on CRAN.

In 2003 Roger Bivand published an extended review of spatial packages. It proposed a class system to support the "data objects offered by GDAL", including 'fundamental' point, line, polygon, and raster types. Furthermore, it suggested interfaces to external libraries should form the basis of modular R packages (Bivand, 2003). To a large extent these ideas were realized in the packages **rgdal** and **sp**. These provided a foundation for spatial data analysis with R, as described in *Applied Spatial Data Analysis with R* (ASDAR) (Bivand et al., 2013), first published in 2008. Ten years later, R's spatial capabilities have evolved substantially but they still build on ideas set-out by Bivand (2003): interfaces to GDAL and PROJ, for example, still power R's high-performance geographic data I/O and CRS transformation capabilities (see Chapters 6 and 7, respectively).

rgdal, released in 2003, provided GDAL bindings for R which greatly enhanced its ability to import data from previously unavailable geographic data formats. The initial release supported only raster drivers but subsequent enhancements provided support for coordinate reference systems (via the PROJ library), reprojections and import of vector file formats (see Chapter 7 for more on file formats). Many of these additional capabilities were developed by Barry Rowlingson and released in the **rgdal** codebase in 2006 (see Rowlingson et al., 2003, and the R-help[27] email list for context).

sp, released in 2005, overcame R's inability to distinguish spatial and non-spatial objects (Pebesma and Bivand, 2005). **sp** grew from a workshop[28] in Vienna in 2003 and was hosted at sourceforge before migrating to R-Forge[29]. Prior to 2005, geographic coordinates were generally treated like any other number. **sp** changed this with its classes and generic methods supporting points, lines, polygons and grids, and attribute data.

sp stores information such as bounding box, coordinate reference system and attributes in slots in `Spatial` objects using the S4 class system, enabling data operations to work on geographic data (see Section 2.2.2). Further, **sp**

---

[26]http://shapelib.maptools.org/

[27]https://stat.ethz.ch/pipermail/r-help/2003-January/028413.html

[28]http://spatial.nhh.no/meetings/vienna/index.html

[29]https://r-forge.r-project.org

provides generic methods such as `summary()` and `plot()` for geographic data. In the following decade, **sp** classes rapidly became popular for geographic data in R and the number of packages that depended on it increased from around 20 in 2008 to over 100 in 2013 (Bivand et al., 2013). As of 2018 almost 500 packages rely on **sp**, making it an important part of the R ecosystem. Prominent R packages using **sp** include: **gstat**, for spatial and spatio-temporal geostatistics; **geosphere**, for spherical trigonometry; and **adehabitat** used for the analysis of habitat selection by animals (Pebesma and Graeler, 2018; Calenge, 2006; Hijmans, 2016).

While **rgdal** and **sp** solved many spatial issues, R still lacked the ability to do geometric operations (see Chapter 5). Colin Rundel addressed this issue by developing **rgeos**, an R interface to the open-source geometry library (GEOS) during a Google Summer of Code project in 2010 (Bivand and Rundel, 2018). **rgeos** enabled GEOS to manipulate **sp** objects, with functions such as `gIntersection()`.

Another limitation of **sp** — its limited support for raster data — was overcome by **raster**, first released in 2010 (Hijmans, 2017). Its class system and functions support a range of raster operations as outlined in Section 2.3. A key feature of **raster** is its ability to work with datasets that are too large to fit into RAM (R's interface to PostGIS supports off-disc operations on vector geographic data). **raster** also supports map algebra (see Section 4.3.2).

In parallel with these developments of class systems and methods came the support for R as an interface to dedicated GIS software. **GRASS** (Bivand, 2000) and follow-on packages **spgrass6** and **rgrass7** (for GRASS GIS 6 and 7, respectively) were prominent examples in this direction (Bivand, 2016a,b). Other examples of bridges between R and GIS include **RSAGA** (Brenning et al., 2018, first published in 2008), **RPyGeo** (Brenning, 2012a, first published in 2008), and **RQGIS** (Muenchow et al., 2017, first published in 2016) (see Chapter 9).

Visualization was not a focus initially, with the bulk of R-spatial development focused on analysis and geographic operations. **sp** provided methods for map making using both the base and lattice plotting system but demand was growing for advanced map making capabilities, especially after the release of **ggplot2** in 2007. **ggmap** extended **ggplot2**'s spatial capabilities (Kahle and Wickham, 2013), by facilitating access to 'basemap' tiles from online services such as Google Maps. Though **ggmap** facilitated map-making with **ggplot2**, its utility was limited by the need to `fortify` spatial objects, which means converting them into long data frames. While this works well for points it is computationally inefficient for lines and polygons, since each coordinate (vertex) is converted into a row, leading to huge data frames to represent complex geometries. Although geographic visualization tended to focus on vector data, raster visualization is supported in **raster** and received a boost with the release of **rasterVis**, which is described in a book on the subject of spatial and temporal data visualization (Lamigueiro, 2018). As of 2018 map making in R is a hot topic with dedicated packages such as **tmap**, **leaflet** and **mapview** all supporting the class system provided by **sf**, the focus of the next chapter (see Chapter 8 for more on visualization).

1.6 Exercises

1. Think about the terms 'GIS', 'GDS' and 'geocomputation' described above. Which (if any) best describes the work you would like to do using geo* methods and software and why?

2. Provide three reasons for using a scriptable language such as R for geocomputation instead of using an established GIS program such as QGIS.

3. Name two advantages and two disadvantages of using mature vs recent packages for geographic data analysis (for example **sp** vs **sf**).

Part I

Foundations

2

Geographic data in R

Prerequisites

This is the first practical chapter of the book, and therefore it comes with some software requirements. We assume that you have an up-to-date version of R installed and that you are comfortable using software with a command-line interface such as the integrated development environment (IDE) RStudio.

If you are new to R, we recommend reading Chapter 2 of the online book *Efficient R Programming* by Gillespie and Lovelace (2016) and learning the basics of the language with reference to resources such as Grolemund and Wickham (2016) or DataCamp[1] before proceeding. Organize your work (e.g., with RStudio projects) and give scripts sensible names such as `chapter-02.R` to document the code you write as you learn.

The packages used in this chapter can be installed with the following commands:[2]

```
install.packages("sf")
install.packages("raster")
install.packages("spData")
devtools::install_github("Nowosad/spDataLarge")
```

If you're running Mac or Linux, the previous command to install **sf** may not work first time. These operating systems (OSs) have 'systems requirements' that are described in the package's README[3]. Various OS-specific instructions can be found online, such as the article *Installation of R 3.5 on Ubuntu 18.04* on the blog rtask.thinkr.fr[4].

---

[1] https://www.datacamp.com/courses/free-introduction-to-r
[2] **spDataLarge** is not on CRAN, meaning it must be installed via **devtools** or with the following command: `install.packages("spDataLarge", repos = "https://nowosad.github.io/drat/", type = "source")`.

All the packages needed to reproduce the contents of the book can be installed
with the following command:

`devtools::install_github("geocompr/geocompkg")`. The necessary packages can be
'loaded' (technically they are attached) with the `library()` function as follows:

```
library(sf)            # classes and functions for vector data
library(raster)        # classes and functions for raster data
```

The other packages that were installed contain data that will be used in the
book:

```
library(spData)        # load geographic data
library(spDataLarge)   # load larger geographic data
```

2.1   Introduction

This chapter will provide brief explanations of the fundamental geographic
data models: vector and raster. We will introduce the theory behind each data
model and the disciplines in which they predominate, before demonstrating
their implementation in R.

The *vector data model* represents the world using points, lines and polygons.
These have discrete, well-defined borders, meaning that vector datasets usually
have a high level of precision (but not necessarily accuracy as we will see in
Section 2.5). The *raster data model* divides the surface up into cells of constant
size. Raster datasets are the basis of background images used in web-mapping
and have been a vital source of geographic data since the origins of aerial
photography and satellite-based remote sensing devices. Rasters aggregate
spatially specific features to a given resolution, meaning that they are consistent
over space and scalable (many worldwide raster datasets are available).

Which to use? The answer likely depends on your domain of application:

- Vector data tends to dominate the social sciences because human settlements
  tend to have discrete borders.
- Raster often dominates in environmental sciences because of the reliance on
  remote sensing data.

There is much overlap in some fields and raster and vector datasets can be
used together: ecologists and demographers, for example, commonly use both
vector and raster data. Furthermore, it is possible to convert between the two
forms (see Section 5.4). Whether your work involves more use of vector or
raster datasets, it is worth understanding the underlying data model before

using them, as discussed in subsequent chapters. This book uses **sf** and **raster** packages to work with vector data and raster datasets, respectively.

2.2 Vector data

Take care when using the word 'vector' as it can have two meanings in this book: geographic vector data and the `vector` class (note the `monospace` font) in R. The former is a data model, the latter is an R class just like `data.frame` and `matrix`. Still, there is a link between the two: the spatial coordinates which are at the heart of the geographic vector data model can be represented in R using `vector` objects.

The geographic vector data model is based on points located within a coordinate reference system (CRS). Points can represent self-standing features (e.g., the location of a bus stop) or they can be linked together to form more complex geometries such as lines and polygons. Most point geometries contain only two dimensions (3-dimensional CRSs contain an additional z value, typically representing height above sea level).

In this system London, for example, can be represented by the coordinates `c(-0.1, 51.5)`. This means that its location is -0.1 degrees east and 51.5 degrees north of the origin. The origin in this case is at 0 degrees longitude (the Prime Meridian) and 0 degree latitude (the Equator) in a geographic ('lon/lat') CRS (Figure 2.1, left panel). The same point could also be approximated in a projected CRS with 'Easting/Northing' values of `c(530000, 180000)` in the British National Grid[5], meaning that London is located 530 km *East* and 180 km *North* of the *origin* of the CRS. This can be verified visually: slightly more than 5 'boxes' — square areas bounded by the gray grid lines 100 km in width — separate the point representing London from the origin (Figure 2.1, right panel).

The location of National Grid's origin, in the sea beyond South West Peninsular, ensures that most locations in the UK have positive Easting and Northing values.[6] There is more to CRSs, as described in Sections 2.4 and 6 but, for the purposes of this section, it is sufficient to know that coordinates consist of two numbers representing distance from an origin, usually in x then y dimensions.

sf is a package providing a class system for geographic vector data. Not only

---

[5]https://en.wikipedia.org/wiki/Ordnance_Survey_National_Grid

[6]The origin we are referring to, depicted in blue in Figure 2.1, is in fact the 'false' origin. The 'true' origin, the location at which distortions are at a minimum, is located at 2° W and 49° N. This was selected by the Ordnance Survey to be roughly in the center of the British landmass longitudinally.

FIGURE 2.1: Illustration of vector (point) data in which location of London (the red X) is represented with reference to an origin (the blue circle). The left plot represents a geographic CRS with an origin at 0° longitude and latitude. The right plot represents a projected CRS with an origin located in the sea west of the South West Peninsula.

does **sf** supersede **sp**, it also provides a consistent command-line interface to GEOS and GDAL, superseding **rgeos** and **rgdal** (described in Section 1.5). This section introduces **sf** classes in preparation for subsequent chapters (Chapters 5 and 7 cover the GEOS and GDAL interface, respectively).

2.2.1 An introduction to simple features

Simple features is an open standard[7] developed and endorsed by the Open Geospatial Consortium (OGC), a not-for-profit organization whose activities we will revisit in a later chapter (in Section 7.5). Simple Features is a hierarchical data model that represents a wide range of geometry types. Of 17 geometry types supported by the specification, only 7 are used in the vast majority of geographic research (see Figure 2.2); these core geometry types are fully supported by the R package **sf** (Pebesma, 2018).[8]

sf can represent all common vector geometry types (raster data classes are not supported by **sf**): points, lines, polygons and their respective 'multi' versions (which group together features of the same type into a single feature). **sf** also supports geometry collections, which can contain multiple geometry types in

---

[7]http://portal.opengeospatial.org/files/?artifact_id=25355

[8]The full OGC standard includes rather exotic geometry types including 'surface' and 'curve' geometry types, which currently have limited application in real world applications. All 17 types can be represented with the **sf** package, although (as of summer 2018) plotting only works for the 'core 7'.

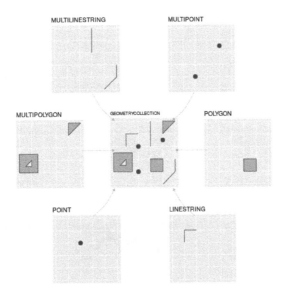

FIGURE 2.2: Simple feature types fully supported by sf.

a single object. **sf** largely supersedes the **sp** ecosystem, which comprises **sp** (Pebesma and Bivand, 2018), **rgdal** for data read/write (Bivand et al., 2018) and **rgeos** for spatial operations (Bivand and Rundel, 2018). The package is well documented, as can be seen on its website and in 6 vignettes, which can be loaded as follows:

```
vignette(package = "sf") # see which vignettes are available
vignette("sf1")          # an introduction to the package
```

As the first vignette explains, simple feature objects in R are stored in a data frame, with geographic data occupying a special column, usually named 'geom' or 'geometry'. We will use the `world` dataset provided by the **spData**, loaded at the beginning of this chapter (see nowosad.github.io/spData[9] for a list of datasets loaded by the package). `world` is a spatial object containing spatial and attribute columns, the names of which are returned by the function `names()` (the last column contains the geographic information):

```
names(world)
#>  [1] "iso_a2"    "name_long" "continent" "region_un" "subregion"
#>  [6] "type"      "area_km2"  "pop"       "lifeExp"   "gdpPercap"
#> [11] "geom"
```

---

[9]https://nowosad.github.io/spData/

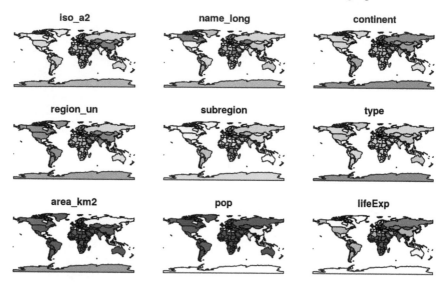

FIGURE 2.3: A spatial plot of the world using the sf package, with a facet for each attribute.

The contents of this `geom` column give `sf` objects their spatial powers: `world$geom` is a 'list column[10]' that contains all the coordinates of the country polygons. The **sf** package provides a `plot()` method for visualizing geographic data: the following command creates Figure 2.3.

```
plot(world)
```

Note that instead of creating a single map, as most GIS programs would, the `plot()` command has created multiple maps, one for each variable in the `world` datasets. This behavior can be useful for exploring the spatial distribution of different variables and is discussed further in Section 2.2.3 below.

Being able to treat spatial objects as regular data frames with spatial powers has many advantages, especially if you are already used to working with data frames. The commonly used `summary()` function, for example, provides a useful overview of the variables within the `world` object.

```
summary(world["lifeExp"])
#>      lifeExp                 geom
#>  Min.   :50.6   MULTIPOLYGON :177
#>  1st Qu.:65.0   epsg:4326    :  0
```

---

[10]https://jennybc.github.io/purrr-tutorial/ls13_list-columns.html

```
#>   Median :72.9     +proj=long...:   0
#>   Mean    :70.9
#>   3rd Qu.:76.8
#>   Max.    :83.6
#>   NA's    :10
```

Although we have only selected one variable for the summary command, it also outputs a report on the geometry. This demonstrates the 'sticky' behavior of the geometry columns of **sf** objects, meaning the geometry is kept unless the user deliberately removes them, as we'll see in Section 3.2. The result provides a quick summary of both the non-spatial and spatial data contained in world: the mean average life expectancy is 71 years (ranging from less than 51 to more than 83 years with a median of 73 years) across all countries.

The word MULTIPOLYGON in the summary output above refers to the geometry type of features (countries) in the world object. This representation is necessary for countries with islands such as Indonesia and Greece. Other geometry types are described in Section 2.2.5.

It is worth taking a deeper look at the basic behavior and contents of this simple feature object, which can usefully be thought of as a 'spatial data frame'.

sf objects are easy to subset. The code below shows its first two rows and three columns. The output shows two major differences compared with a regular data.frame: the inclusion of additional geographic data (geometry type, dimension, bbox and CRS information - epsg (SRID), proj4string), and the presence of a geometry column, here named geom:

```
world_mini = world[1:2, 1:3]
world_mini
#> Simple feature collection with 2 features and 3 fields
#> geometry type:    MULTIPOLYGON
#> dimension:        XY
#> bbox:             xmin: -180 ymin: -18.3 xmax: 180 ymax: -0.95
#> epsg (SRID):      4326
#> proj4string:      +proj=longlat +datum=WGS84 +no_defs
#>   iso_a2 name_long continent                              geom
#> 1     FJ      Fiji   Oceania MULTIPOLYGON (((180 -16.1, ...
#> 2     TZ  Tanzania    Africa MULTIPOLYGON (((33.9 -0.95,...
```

All this may seem rather complex, especially for a class system that is supposed to be simple. However, there are good reasons for organizing things this way and using **sf**.

Before describing each geometry type that the **sf** package supports, it is worth taking a step back to understand the building blocks of sf objects. Section 2.2.8 shows how simple features objects are data frames, with special geometry columns. These spatial columns are often called geom or geometry: world$geom refers to the spatial element of the world object described above. These geometry columns are 'list columns' of class sfc (see Section 2.2.7). In turn, sfc objects are composed of one or more objects of class sfg: simple feature geometries that we describe in Section 2.2.6.

To understand how the spatial components of simple features work, it is vital to understand simple feature geometries. For this reason we cover each currently supported simple features geometry type in Section 2.2.5 before moving on to describe how these can be represented in R using sfg objects, which form the basis of sfc and eventually full sf objects.

The preceding code chunk uses = to create a new object called world_mini in the command world_mini = world[1:2, 1:3]. This is called assignment. An equivalent command to achieve the same result is world_mini <- world[1:2, 1:3]. Although 'arrow assigment' is more commonly used, we use 'equals assignment' because it's slightly faster to type and easier to teach due to compatibility with commonly used languages such as Python and JavaScript. Which to use is largely a matter of preference as long as you're consistent (packages such as **styler** can be used to change style).

2.2.2 Why simple features?

Simple features is a widely supported data model that underlies data structures in many GIS applications including QGIS and PostGIS. A major advantage of this is that using the data model ensures your work is cross-transferable to other set-ups, for example importing from and exporting to spatial databases.

A more specific question from an R perspective is "why use the **sf** package when **sp** is already tried and tested"? There are many reasons (linked to the advantages of the simple features model) including:

- Fast reading and writing of data.
- Enhanced plotting performance.
- **sf** objects can be treated as data frames in most operations.
- **sf** functions can be combined using %>% operator and works well with the tidyverse[11] collection of R packages.
- **sf** function names are relatively consistent and intuitive (all begin with st_).

Due to such advantages, some spatial packages (including **tmap**, **mapview** and **tidycensus**) have added support for **sf**. However, it will take many years

---

[11]http://tidyverse.org/

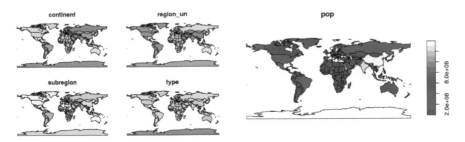

FIGURE 2.4: Plotting with sf, with multiple variables (left) and a single variable (right).

for most packages to transition and some will never switch. Fortunately, these can still be used in a workflow based on sf objects, by converting them to the Spatial class used in **sp**:

```
library(sp)
world_sp = as(world, Class = "Spatial")
# sp functions ...
```

Spatial objects can be converted back to sf in the same way or with st_as_sf():

```
world_sf = st_as_sf(world_sp, "sf")
```

2.2.3   Basic map making

Basic maps are created in **sf** with plot(). By default this creates a multi-panel plot (like **sp**'s spplot()), one sub-plot for each variable of the object, as illustrated in the left-hand panel in Figure 2.4. A legend or 'key' with a continuous color is produced if the object to be plotted has a single variable (see the right-hand panel). Colors can also be set with col =, although this will not create a continuous palette or a legend.

```
plot(world[3:6])
plot(world["pop"])
```

Plots are added as layers to existing images by setting add = TRUE.[12] To demonstrate this, and to provide a taster of content covered in Chapters 3 and 4 on

---

[12]plot()ing of **sf** objects uses sf:::plot.sf() behind the scenes. plot() is a generic method that behaves differently depending on the class of object being plotted.

attribute and spatial data operations, the subsequent code chunk combines
countries in Asia:

```
world_asia = world[world$continent == "Asia", ]
asia = st_union(world_asia)
```

We can now plot the Asian continent over a map of the world. Note that the
first plot must only have one facet for add = TRUE to work. If the first plot has
a key, reset = FALSE must be used (result not shown):

```
plot(world["pop"], reset = FALSE)
plot(asia, add = TRUE, col = "red")
```

Adding layers in this way can be used to verify the geographic correspondence
between layers: the plot() function is fast to execute and requires few lines
of code, but does not create interactive maps with a wide range of options.
For more advanced map making we recommend using dedicated visualization
packages such as **tmap** (see Chapter 8).

2.2.4   Base plot arguments

There are various ways to modify maps with sf's plot() method. Because **sf**
extends base R plotting methods plot()'s arguments such as main = (which
specifies the title of the map) work with sf objects (see ?graphics::plot and
?par).[13]

Figure 2.5 illustrates this flexibility by overlaying circles, whose diameters (set
with cex =) represent country populations, on a map of the world. A basic
version of the map can be created with the following commands (see exercises
at the end of this chapter and the script 02-contplot.R[14] to create Figure 2.5):

```
plot(world["continent"], reset = FALSE)
cex = sqrt(world$pop) / 10000
world_cents = st_centroid(world, of_largest = TRUE)
plot(st_geometry(world_cents), add = TRUE, cex = cex)
```

The code above uses the function st_centroid() to convert one geometry type
(polygons) to another (points) (see Chapter 5), the aesthetics of which are
varied with the cex argument.

sf's plot method also has arguments specific to geographic data. expandBB, for
example, can be used plot an sf object in context: it takes a numeric vector

---

[13]Note: many plot arguments are ignored in facet maps, when more than one sf column is
plotted.

[14]https://github.com/Robinlovelace/geocompr/blob/master/code/02-contpop.R

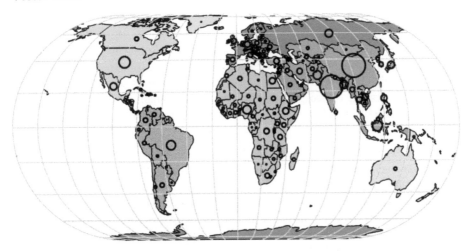

FIGURE 2.5: Country continents (represented by fill color) and 2015 populations (represented by circles, with area proportional to population).

FIGURE 2.6: India in context, demonstrating the expandBB argument.

of length four that expands the bounding box of the plot relative to zero in the following order: bottom, left, top, right. This is used to plot India in the context of its giant Asian neighbors, with an emphasis on China to the east, in the following code chunk, which generates Figure 2.6 (see exercises below on adding text to plots):

```
india = world[world$name_long == "India", ]
plot(st_geometry(india), expandBB = c(0, 0.2, 0.1, 1), col = "gray", lwd = 3)
plot(world_asia[0], add = TRUE)
```

Note the use of [0] to keep only the geometry column and lwd to emphasize India. See Section 8.6 for other visualization techniques for representing a range of geometry types, the subject of the next section.

2.2.5 Geometry types

Geometries are the basic building blocks of simple features. Simple features in R can take on one of the 17 geometry types supported by the **sf** package. In this chapter we will focus on the seven most commonly used types: POINT, LINESTRING, POLYGON, MULTIPOINT, MULTILINESTRING, MULTIPOLYGON and GEOME-TRYCOLLECTION. Find the whole list of possible feature types in the PostGIS manual[15].

Generally, well-known binary (WKB) or well-known text (WKT) are the standard encoding for simple feature geometries. WKB representations are usually hexadecimal strings easily readable for computers. This is why GIS and spatial databases use WKB to transfer and store geometry objects. WKT, on the other hand, is a human-readable text markup description of simple features. Both formats are exchangeable, and if we present one, we will naturally choose the WKT representation.

The basis for each geometry type is the point. A point is simply a coordinate in 2D, 3D or 4D space (see vignette("sf1") for more information) such as (see left panel in Figure 2.7):

• POINT (5 2)

A linestring is a sequence of points with a straight line connecting the points, for example (see middle panel in Figure 2.7):

• LINESTRING (1 5, 4 4, 4 1, 2 2, 3 2)

A polygon is a sequence of points that form a closed, non-intersecting ring. Closed means that the first and the last point of a polygon have the same coordinates (see right panel in Figure 2.7).[16]

• Polygon without a hole: POLYGON ((1 5, 2 2, 4 1, 4 4, 1 5))

So far we have created geometries with only one geometric entity per feature. However, **sf** also allows multiple geometries to exist within a single feature (hence the term 'geometry collection') using "multi" version of each geometry type:

• Multipoint: MULTIPOINT (5 2, 1 3, 3 4, 3 2)
• Multilinestring: MULTILINESTRING ((1 5, 4 4, 4 1, 2 2, 3 2), (1 2, 2 4))
• Multipolygon: MULTIPOLYGON (((1 5, 2 2, 4 1, 4 4, 1 5), (0 2, 1 2, 1 3, 0 3, 0 2)))

Finally, a geometry collection can contain any combination of geometries including (multi)points and linestrings (see Figure 2.9):

---

[15] http://postgis.net/docs/using_postgis_dbmanagement.html

[16] By definition, a polygon has one exterior boundary (outer ring) and can have zero or more interior boundaries (inner rings), also known as holes. A polygon with a hole would be, for example, POLYGON ((1 5, 2 2, 4 1, 4 4, 1 5), (2 4, 3 4, 3 3, 2 3, 2 4))

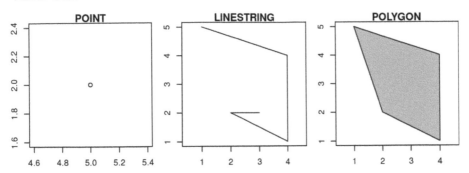

FIGURE 2.7: Illustration of point, linestring and polygon geometries.

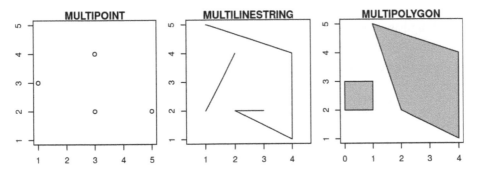

FIGURE 2.8: Illustration of multi* geometries.

- Geometry collection: GEOMETRYCOLLECTION (MULTIPOINT (5 2, 1 3, 3 4, 3 2), LINESTRING (1 5, 4 4, 4 1, 2 2, 3 2))

2.2.6 Simple feature geometries (sfg)

The sfg class represents the different simple feature geometry types in R: point, linestring, polygon (and their 'multi' equivalents, such as multipoints) or geometry collection.

Usually you are spared the tedious task of creating geometries on your own since you can simply import an already existing spatial file. However, there are a set of functions to create simple feature geometry objects (sfg) from scratch if needed. The names of these functions are simple and consistent, as they all start with the st_ prefix and end with the name of the geometry type in lowercase letters:

- A point: st_point()
- A linestring: st_linestring()
- A polygon: st_polygon()
- A multipoint: st_multipoint()
- A multilinestring: st_multilinestring()

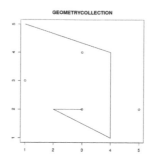

FIGURE 2.9: Illustration of a geometry collection.

- A multipolygon: `st_multipolygon()`
- A geometry collection: `st_geometrycollection()`

`sfg` objects can be created from three base R data types:

1. A numeric vector: a single point
2. A matrix: a set of points, where each row represents a point, a multipoint or linestring
3. A list: a collection of objects such as matrices, multilinestrings or geometry collections

The function `st_point()` creates single points from numeric vectors:

```
st_point(c(5, 2))                # XY point
#> POINT (5 2)
st_point(c(5, 2, 3))             # XYZ point
#> POINT Z (5 2 3)
st_point(c(5, 2, 1), dim = "XYM") # XYM point
#> POINT M (5 2 1)
st_point(c(5, 2, 3, 1))          # XYZM point
#> POINT ZM (5 2 3 1)
```

The results show that XY (2D coordinates), XYZ (3D coordinates) and XYZM (3D with an additional variable, typically measurement accuracy) point types are created from vectors of length 2, 3, and 4, respectively. The XYM type must be specified using the `dim` argument (which is short for dimension).

By contrast, use matrices in the case of multipoint (`st_multipoint()`) and linestring (`st_linestring()`) objects:

```
# the rbind function simplifies the creation of matrices
## MULTIPOINT
multipoint_matrix = rbind(c(5, 2), c(1, 3), c(3, 4), c(3, 2))
st_multipoint(multipoint_matrix)
#> MULTIPOINT (5 2, 1 3, 3 4, 3 2)
## LINESTRING
linestring_matrix = rbind(c(1, 5), c(4, 4), c(4, 1), c(2, 2), c(3, 2))
st_linestring(linestring_matrix)
#> LINESTRING (1 5, 4 4, 4 1, 2 2, 3 2)
```

Finally, use lists for the creation of multilinestrings, (multi-)polygons and geometry collections:

```
## POLYGON
polygon_list = list(rbind(c(1, 5), c(2, 2), c(4, 1), c(4, 4), c(1, 5)))
st_polygon(polygon_list)
#> POLYGON ((1 5, 2 2, 4 1, 4 4, 1 5))


## POLYGON with a hole
polygon_border = rbind(c(1, 5), c(2, 2), c(4, 1), c(4, 4), c(1, 5))
polygon_hole = rbind(c(2, 4), c(3, 4), c(3, 3), c(2, 3), c(2, 4))
polygon_with_hole_list = list(polygon_border, polygon_hole)
st_polygon(polygon_with_hole_list)
#> POLYGON ((1 5, 2 2, 4 1, 4 4, 1 5), (2 4, 3 4, 3 3, 2 3, 2 4))


## MULTILINESTRING
multilinestring_list = list(rbind(c(1, 5), c(4, 4), c(4, 1), c(2, 2), c(3, 2)),
                            rbind(c(1, 2), c(2, 4)))
st_multilinestring((multilinestring_list))
#> MULTILINESTRING ((1 5, 4 4, 4 1, 2 2, 3 2), (1 2, 2 4))


## MULTIPOLYGON
multipolygon_list = list(list(rbind(c(1, 5), c(2, 2), c(4, 1), c(4, 4), c(1, 5))),
                         list(rbind(c(0, 2), c(1, 2), c(1, 3), c(0, 3), c(0, 2))))
st_multipolygon(multipolygon_list)
#> MULTIPOLYGON (((1 5, 2 2, 4 1, 4 4, 1 5)), ((0 2, 1 2, 1 3, 0 3, 0 2)))


## GEOMETRYCOLLECTION
gemetrycollection_list = list(st_multipoint(multipoint_matrix),
                              st_linestring(linestring_matrix))
```

```
st_geometrycollection(gemetrycollection_list)
#> GEOMETRYCOLLECTION (MULTIPOINT (5 2, 1 3, 3 4, 3 2),
#>   LINESTRING (1 5, 4 4, 4 1, 2 2, 3 2))
```

2.2.7　Simple feature columns (sfc)

One `sfg` object contains only a single simple feature geometry. A simple feature geometry column (`sfc`) is a list of `sfg` objects, which is additionally able to contain information about the coordinate reference system in use. For instance, to combine two simple features into one object with two features, we can use the `st_sfc()` function. This is important since `sfc` represents the geometry column in **sf** data frames:

```
# sfc POINT
point1 = st_point(c(5, 2))
point2 = st_point(c(1, 3))
points_sfc = st_sfc(point1, point2)
points_sfc
#> Geometry set for 2 features
#> geometry type:  POINT
#> dimension:      XY
#> bbox:           xmin: 1 ymin: 2 xmax: 5 ymax: 3
#> epsg (SRID):    NA
#> proj4string:    NA
#> POINT (5 2)
#> POINT (1 3)
```

In most cases, an `sfc` object contains objects of the same geometry type. Therefore, when we convert `sfg` objects of type polygon into a simple feature geometry column, we would also end up with an `sfc` object of type polygon, which can be verified with `st_geometry_type()`. Equally, a geometry column of multilinestrings would result in an `sfc` object of type multilinestring:

```
# sfc POLYGON
polygon_list1 = list(rbind(c(1, 5), c(2, 2), c(4, 1), c(4, 4), c(1, 5)))
polygon1 = st_polygon(polygon_list1)
polygon_list2 = list(rbind(c(0, 2), c(1, 2), c(1, 3), c(0, 3), c(0, 2)))
polygon2 = st_polygon(polygon_list2)
polygon_sfc = st_sfc(polygon1, polygon2)
st_geometry_type(polygon_sfc)
#> [1] POLYGON POLYGON
#> 18 Levels: GEOMETRY POINT LINESTRING POLYGON ... TRIANGLE
```

```
# sfc MULTILINESTRING
multilinestring_list1 = list(rbind(c(1, 5), c(4, 4), c(4, 1), c(2, 2), c(3, 2)),
                             rbind(c(1, 2), c(2, 4)))
multilinestring1 = st_multilinestring((multilinestring_list1))
multilinestring_list2 = list(rbind(c(2, 9), c(7, 9), c(5, 6), c(4, 7), c(2, 7)),
                             rbind(c(1, 7), c(3, 8)))
multilinestring2 = st_multilinestring((multilinestring_list2))
multilinestring_sfc = st_sfc(multilinestring1, multilinestring2)
st_geometry_type(multilinestring_sfc)
#> [1] MULTILINESTRING MULTILINESTRING
#> 18 Levels: GEOMETRY POINT LINESTRING POLYGON ... TRIANGLE
```

It is also possible to create an `sfc` object from `sfg` objects with different geometry types:

```
# sfc GEOMETRY
point_multilinestring_sfc = st_sfc(point1, multilinestring1)
st_geometry_type(point_multilinestring_sfc)
#> [1] POINT           MULTILINESTRING
#> 18 Levels: GEOMETRY POINT LINESTRING POLYGON ... TRIANGLE
```

As mentioned before, `sfc` objects can additionally store information on the coordinate reference systems (CRS). To specify a certain CRS, we can use the `epsg` (SRID) or `proj4string` attributes of an `sfc` object. The default value of `epsg` (SRID) and `proj4string` is NA (*Not Available*), as can be verified with `st_crs()`:

```
st_crs(points_sfc)
#> Coordinate Reference System: NA
```

All geometries in an `sfc` object must have the same CRS. We can add coordinate reference system as a `crs` argument of `st_sfc()`. This argument accepts an integer with the `epsg` code such as 4326, which automatically adds the 'proj4string' (see Section 2.4):

```
# EPSG definition
points_sfc_wgs = st_sfc(point1, point2, crs = 4326)
st_crs(points_sfc_wgs)
#> Coordinate Reference System:
#>   EPSG: 4326
#>   proj4string: "+proj=longlat +datum=WGS84 +no_defs"
```

It also accepts a raw proj4string (result not shown):

```
# PROJ4STRING definition
st_sfc(point1, point2, crs = "+proj=longlat +datum=WGS84 +no_defs")
```

 Sometimes `st_crs()` will return a `proj4string` but not an `epsg` code. This is because there is no general method to convert from `proj4string` to `epsg` (see Chapter 6).

2.2.8   The sf class

Sections 2.2.5 to 2.2.7 deal with purely geometric objects, 'sf geometry' and 'sf column' objects, respectively. These are geographic building blocks of geographic vector data represented as simple features. The final building block is non-geographic attributes, representing the name of the feature or other attributes such as measured values, groups, and other things.

To illustrate attributes, we will represent a temperature of 25°C in London on June 21st, 2017. This example contains a geometry (the coordinates), and three attributes with three different classes (place name, temperature and date).[17] Objects of class `sf` represent such data by combining the attributes (`data.frame`) with the simple feature geometry column (`sfc`). They are created with `st_sf()` as illustrated below, which creates the London example described above:

```
lnd_point = st_point(c(0.1, 51.5))          # sfg object
lnd_geom = st_sfc(lnd_point, crs = 4326)    # sfc object
lnd_attrib = data.frame(                    # data.frame object
  name = "London",
  temperature = 25,
  date = as.Date("2017-06-21")
  )
lnd_sf = st_sf(lnd_attrib, geometry = lnd_geom)   # sf object
```

What just happened? First, the coordinates were used to create the simple feature geometry (`sfg`). Second, the geometry was converted into a simple feature geometry column (`sfc`), with a CRS. Third, attributes were stored in a `data.frame`, which was combined with the `sfc` object with `st_sf()`. This results in an `sf` object, as demonstrated below (some output is ommited):

---

[17]Other attributes might include an urbanity category (city or village), or a remark if the measurement was made using an automatic station.

```
lnd_sf
#> Simple feature collection with 1 features and 3 fields
#> ...
#>     name temperature       date        geometry
#> 1 London          25 2017-06-21 POINT (0.1 51.5)
```

```
class(lnd_sf)
#> [1] "sf"         "data.frame"
```

The result shows that `sf` objects actually have two classes, `sf` and `data.frame`. Simple features are simply data frames (square tables), but with spatial attributes stored in a list column, usually called `geometry`, as described in Section 2.2.1. This duality is central to the concept of simple features: most of the time a `sf` can be treated as and behaves like a `data.frame`. Simple features are, in essence, data frames with a spatial extension.

2.3 Raster data

The geographic raster data model usually consists of a raster header and a matrix (with rows and columns) representing equally spaced cells (often also called pixels; Figure 2.10:A).[18] The raster header defines the coordinate reference system, the extent and the origin. The origin (or starting point) is frequently the coordinate of the lower-left corner of the matrix (the **raster** package, however, uses the upper left corner, by default (Figure 2.10:B)). The header defines the extent via the number of columns, the number of rows and the cell size resolution. Hence, starting from the origin, we can easily access and modify each single cell by either using the ID of a cell (Figure 2.10:B) or by explicitly specifying the rows and columns. This matrix representation avoids storing explicitly the coordinates for the four corner points (in fact it only stores one coordinate, namely the origin) of each cell corner as would be the case for rectangular vector polygons. This and map algebra makes raster processing much more efficient and faster than vector data processing. However, in contrast to vector data, the cell of one raster layer can only hold a single value. The value might be numeric or categorical (Figure 2.10:C).

Raster maps usually represent continuous phenomena such as elevation, temperature, population density or spectral data (Figure 2.11). Of course, we can

---

[18]Depending on the file format the header is part of the actual image data file, e.g., GeoTIFF, or stored in an extra header or world file, e.g., ASCII grid formats. There is also the headerless (flat) binary raster format which should facilitate the import into various software programs.

A. Cell IDs

1	2	3	4
5	6	7	8
9	10	11	12
13	14	15	16

B. Cell values

22	74	28	91
72	84	NA	85
NA	92	24	53
31	62	56	5

C. Colored values

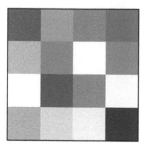

FIGURE 2.10: Raster data types: (A) cell IDs, (B) cell values, (C) a colored raster map.

A. Continuous data

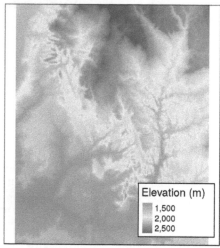

Elevation (m)
1,500
2,000
2,500

B. Categorical data

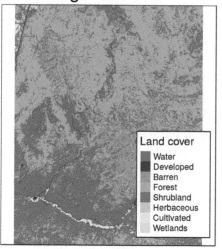

Land cover
Water
Developed
Barren
Forest
Shrubland
Herbaceous
Cultivated
Wetlands

FIGURE 2.11: Examples of continuous and categorical rasters.

represent discrete features such as soil or land-cover classes also with the help of a raster data model (Figure 2.11). Consequently, the discrete borders of these features become blurred, and depending on the spatial task a vector representation might be more suitable.

2.3.1 An introduction to raster

The **raster** package supports raster objects in R. It provides an extensive set of functions to create, read, export, manipulate and process raster datasets. Aside from general raster data manipulation, **raster** provides many low-level functions that can form the basis to develop more advanced raster functionality.

raster also lets you work on large raster datasets that are too large to fit into the main memory. In this case, **raster** provides the possibility to divide the raster into smaller chunks (rows or blocks), and processes these iteratively instead of loading the whole raster file into RAM (for more information, please refer to `vignette("functions", package = "raster")`.

For the illustration of **raster** concepts, we will use datasets from the **spData-Large** (note these packages were loaded at the beginning of the chapter). It consists of a few raster objects and one vector object covering an area of the Zion National Park (Utah, USA). For example, `srtm.tif` is a digital elevation model of this area (for more details, see its documentation `?srtm`). First, let's create a `RasterLayer` object named `new_raster`:

```
raster_filepath = system.file("raster/srtm.tif", package = "spDataLarge")
new_raster = raster(raster_filepath)
```

Typing the name of the raster into the console, will print out the raster header (extent, dimensions, resolution, CRS) and some additional information (class, data source name, summary of the raster values):

```
new_raster
#> class       : RasterLayer
#> dimensions  : 457, 465, 212505  (nrow, ncol, ncell)
#> resolution  : 0.000833, 0.000833  (x, y)
#> extent      : -113, -113, 37.1, 37.5  (xmin, xmax, ymin, ymax)
#> coord. ref. : +proj=longlat +datum=WGS84 +no_defs +ellps=WGS84 +towgs84=0,0,0
#> data source : /home/robin/R/x86_64-pc-linux../3.5/spDataLarge/raster/srtm.tif
#> names       : srtm
#> values      : 1024, 2892  (min, max)
```

Dedicated functions report each component: `dim(new_raster)` returns the number of rows, columns and layers; the `ncell()` function the number of cells (pixels); `res()` the raster's spatial resolution; `extent()` its spatial extent; and `crs()` its coordinate reference system (raster reprojection is covered in Section 6.6). `inMemory()` reports whether the raster data is stored in memory (the default) or on disk.

`help("raster-package")` returns a full list of all available **raster** functions.

2.3.2 Basic map making

Similar to the **sf** package, **raster** also provides `plot()` methods for its own classes.

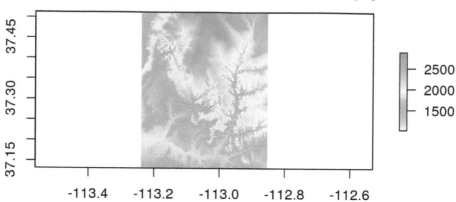

FIGURE 2.12: Basic raster plot.

```
plot(new_raster)
```

There are several other approaches for plotting raster data in R that are outside the scope of this section, including:

- Functions such as spplot() and levelplot() (from the **sp** and **rasterVis** packages, respectively) to create facets, a common technique for visualizing change over time.
- Packages such as **tmap**, **mapview** and **leaflet** to create interactive maps of raster and vector objects (see Chapter 8).

2.3.3 Raster classes

The RasterLayer class represents the simplest form of a raster object, and consists of only one layer. The easiest way to create a raster object in R is to read-in a raster file from disk or from a server.

```
raster_filepath = system.file("raster/srtm.tif", package = "spDataLarge")
new_raster = raster(raster_filepath)
```

The **raster** package supports numerous drivers with the help of **rgdal**. To find out which drivers are available on your system, run raster::writeFormats() and rgdal::gdalDrivers().

Rasters can also be created from scratch using the raster() function. This is illustrated in the subsequent code chunk, which results in a new RasterLayer object. The resulting raster consists of 36 cells (6 columns and 6 rows specified by nrows and ncols) centered around the Prime Meridian and the Equator (see xmn, xmx, ymn and ymx parameters). The CRS is the default of raster objects: WGS84. This means the unit of the resolution is in degrees which we set to

0.5 (`res`). Values (`vals`) are assigned to each cell: 1 to cell 1, 2 to cell 2, and so on. Remember: `raster()` fills cells row-wise (unlike `matrix()`) starting at the upper left corner, meaning the top row contains the values 1 to 6, the second 7 to 12, etc.

```
new_raster2 = raster(nrows = 6, ncols = 6, res = 0.5,
                     xmn = -1.5, xmx = 1.5, ymn = -1.5, ymx = 1.5,
                     vals = 1:36)
```

For other ways of creating raster objects, see `?raster`.

Aside from `RasterLayer`, there are two additional classes: `RasterBrick` and `RasterStack`. Both can handle multiple layers, but differ regarding the number of supported file formats, type of internal representation and processing speed.

A `RasterBrick` consists of multiple layers, which typically correspond to a single multispectral satellite file or a single multilayer object in memory. The `brick()` function creates a `RasterBrick` object. Usually, you provide it with a filename to a multilayer raster file but might also use another raster object and other spatial objects (see `?brick` for all supported formats).

```
multi_raster_file = system.file("raster/landsat.tif", package = "spDataLarge")
r_brick = brick(multi_raster_file)
```

```
r_brick
#> class        : RasterBrick
#> resolution   : 30, 30  (x, y)
#> ...
#> names        : landsat.1, landsat.2, landsat.3, landsat.4
#> min values   :     7550,     6404,     5678,     5252
#> max values   :    19071,    22051,    25780,    31961
```

`nlayers()` retrieves the number of layers stored in a `Raster*` object:

```
nlayers(r_brick)
#> [1] 4
```

A `RasterStack` is similar to a `RasterBrick` in the sense that it consists also of multiple layers. However, in contrast to `RasterBrick`, `RasterStack` allows you to connect several raster objects stored in different files or multiply objects in memory. More specifically, a `RasterStack` is a list of `RasterLayer` objects with the same extent and resolution. Hence, one way to create it is with the help of spatial objects already existing in R's global environment. And again, one can simply specify a path to a file stored on disk.

```
raster_on_disk = raster(r_brick, layer = 1)
raster_in_memory = raster(xmn = 301905, xmx = 335745,
                          ymn = 4111245, ymx = 4154085,
                          res = 30)
values(raster_in_memory) = sample(seq_len(ncell(raster_in_memory)))
crs(raster_in_memory) = crs(raster_on_disk)
```

```
r_stack = stack(raster_in_memory, raster_on_disk)
r_stack
#> class : RasterStack
#> dimensions : 1428, 1128, 1610784, 2
#> resolution : 30, 30
#> ...
#> names       :    layer, landsat.1
#> min values  :        1,      7550
#> max values  : 1610784,     19071
```

Another difference is that the processing time for RasterBrick objects is usually shorter than for RasterStack objects.

Decision on which Raster* class should be used depends mostly on a character of input data. Processing of a single mulitilayer file or object is the most effective with RasterBrick, while RasterStack allows calculations based on many files, many Raster* objects, or both.

 Operations on RasterBrick and RasterStack objects will typically return a RasterBrick.

2.4   Coordinate Reference Systems

Vector and raster spatial data types share concepts intrinsic to spatial data. Perhaps the most fundamental of these is the Coordinate Reference System (CRS), which defines how the spatial elements of the data relate to the surface of the Earth (or other bodies). CRSs are either geographic or projected, as introduced at the beginning of this chapter (see Figure 2.1). This section will explain each type, laying the foundations for Section 6 on CRS transformations.

2.4.1 Geographic coordinate systems

Geographic coordinate systems identify any location on the Earth's surface using two values — longitude and latitude. *Longitude* is location in the East-West direction in angular distance from the Prime Meridian plane. *Latitude* is angular distance North or South of the equatorial plane. Distances in geographic CRSs are therefore not measured in meters. This has important consequences, as demonstrated in Section 6.

The surface of the Earth in geographic coordinate systems is represented by a spherical or ellipsoidal surface. Spherical models assume that the Earth is a perfect sphere of a given radius. Spherical models have the advantage of simplicity but are rarely used because they are inaccurate: the Earth is not a sphere! Ellipsoidal models are defined by two parameters: the equatorial radius and the polar radius. These are suitable because the Earth is compressed: the equatorial radius is around 11.5 km longer than the polar radius (Maling, 1992).[19]

Ellipsoids are part of a wider component of CRSs: the *datum*. This contains information on what ellipsoid to use (with the `ellps` parameter in the PROJ CRS library) and the precise relationship between the Cartesian coordinates and location on the Earth's surface. These additional details are stored in the `towgs84` argument of proj4string[20] notation (see proj4.org/parameters.html[21] for details). These allow local variations in Earth's surface, for example due to large mountain ranges, to be accounted for in a local CRS. There are two types of datum — local and geocentric. In a *local datum* such as NAD83 the ellipsoidal surface is shifted to align with the surface at a particular location. In a *geocentric datum* such as WGS84 the center is the Earth's center of gravity and the accuracy of projections is not optimized for a specific location. Available datum definitions can be seen by executing `st_proj_info(type = "datum")`.

2.4.2 Projected coordinate reference systems

Projected CRSs are based on Cartesian coordinates on an implicitly flat surface. They have an origin, x and y axes, and a linear unit of measurement such as meters. All projected CRSs are based on a geographic CRS, described in the previous section, and rely on map projections to convert the three-dimensional surface of the Earth into Easting and Northing (x and y) values in a projected CRS.

---

[19]The degree of compression is often referred to as *flattening*, defined in terms of the equatorial radius (a) and polar radius (b) as follows: $f = (a - b)/a$. The terms *ellipticity* and *compression* can also be used (Maling, 1992). Because f is a rather small value, digital ellipsoid models use the 'inverse flattening' ($rf = 1/f$) to define the Earth's compression. Values of a and rf in various ellipsoidal models can be seen by executing `st_proj_info(type = "ellps")`.

[20]https://proj4.org/operations/conversions/latlon.html?highlight=towgs#cmdoption-arg-towgs84

[21]https://proj4.org/usage/projections.html

This transition cannot be done without adding some distortion. Therefore, some properties of the Earth's surface are distorted in this process, such as area, direction, distance, and shape. A projected coordinate system can preserve only one or two of those properties. Projections are often named based on a property they preserve: equal-area preserves area, azimuthal preserve direction, equidistant preserve distance, and conformal preserve local shape.

There are three main groups of projection types - conic, cylindrical, and planar. In a conic projection, the Earth's surface is projected onto a cone along a single line of tangency or two lines of tangency. Distortions are minimized along the tangency lines and rise with the distance from those lines in this projection. Therefore, it is the best suited for maps of mid-latitude areas. A cylindrical projection maps the surface onto a cylinder. This projection could also be created by touching the Earth's surface along a single line of tangency or two lines of tangency. Cylindrical projections are used most often when mapping the entire world. A planar projection projects data onto a flat surface touching the globe at a point or along a line of tangency. It is typically used in mapping polar regions. `st_proj_info(type = "proj")` gives a list of the available projections supported by the PROJ library.

2.4.3   CRSs in R

Two main ways to describe CRS in R are an `epsg` code or a `proj4string` definition. Both of these approaches have advantages and disadvantages. An `epsg` code is usually shorter, and therefore easier to remember. The code also refers to only one, well-defined coordinate reference system. On the other hand, a `proj4string` definition allows you more flexibility when it comes to specifying different parameters such as the projection type, the datum and the ellipsoid.[22] This way you can specify many different projections, and modify existing ones. This also makes the `proj4string` approach more complicated. `epsg` points to exactly one particular CRS.

Spatial R packages support a wide range of CRSs and they use the long-established PROJ[23] library. Other than searching for EPSG codes online, another quick way to find out about available CRSs is via the `rgdal::make_EPSG()` function, which outputs a data frame of available projections. Before going into more detail, it's worth learning how to view and filter them inside R, as this could save time trawling the internet. The following code will show available CRSs interactively, allowing you to filter ones of interest (try filtering for the OSGB CRSs for example):

---

[22]A complete list of the `proj4string` parameters can be found at https://proj4.org/.
[23]http://proj4.org/

```
crs_data = rgdal::make_EPSG()
View(crs_data)
```

In **sf** the CRS of an object can be retrieved using `st_crs()`. For this, we need to read-in a vector dataset:

```
vector_filepath = system.file("vector/zion.gpkg", package = "spDataLarge")
new_vector = st_read(vector_filepath)
```

Our new object, `new_vector`, is a polygon representing the borders of Zion National Park (`?zion`).

```
st_crs(new_vector) # get CRS
#> Coordinate Reference System:
#> No EPSG code
#> proj4string: "+proj=utm +zone=12 +ellps=GRS80 ... +units=m +no_defs"
```

In cases when a coordinate reference system (CRS) is missing or the wrong CRS is set, the `st_set_crs()` function can be used:

```
new_vector = st_set_crs(new_vector, 4326) # set CRS
#> Warning: st_crs<- : replacing crs does not reproject data; use st_transform
#> for that
```

The warning message informs us that the `st_set_crs()` function does not transform data from one CRS to another.

The `projection()` function can be used to access CRS information from a `Raster*` object:

```
projection(new_raster) # get CRS
#> [1] "+proj=longlat +datum=WGS84 +no_defs +ellps=WGS84 +towgs84=0,0,0"
```

The same function, `projection()`, is used to set a CRS for raster objects. The main difference, compared to vector data, is that raster objects only accept `proj4` definitions:

```
projection(new_raster) = "+proj=utm +zone=12 +ellps=GRS80 +towgs84=0,0,0,0,0,0,0
                          +units=m +no_defs" # set CRS
```

We will expand on CRSs and how to project from one CRS to another in much more detail in Chapter 6.

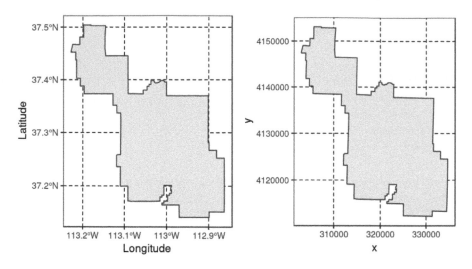

FIGURE 2.13: Examples of geographic (WGS 84; left) and projected (NAD83 / UTM zone 12N; right) coordinate systems for a vector data type.

2.5 Units

An important feature of CRSs is that they contain information about spatial units. Clearly, it is vital to know whether a house's measurements are in feet or meters, and the same applies to maps. It is good cartographic practice to add a *scale bar* onto maps to demonstrate the relationship between distances on the page or screen and distances on the ground. Likewise, it is important to formally specify the units in which the geometry data or pixels are measured to provide context, and ensure that subsequent calculations are done in context.

A novel feature of geometry data in sf objects is that they have *native support* for units. This means that distance, area and other geometric calculations in **sf** return values that come with a units attribute, defined by the **units** package (Pebesma et al., 2016). This is advantageous, preventing confusion caused by different units (most CRSs use meters, some use feet) and providing information on dimensionality. This is demonstrated in the code chunk below, which calculates the area of Luxembourg:

```
luxembourg = world[world$name_long == "Luxembourg", ]
```

```
st_area(luxembourg)
#> 2.42e+09 [m^2]
```

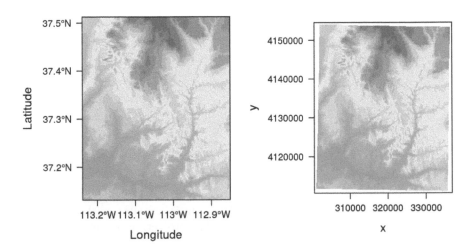

FIGURE 2.14: Examples of geographic (WGS 84; left) and projected (NAD83 / UTM zone 12N; right) coordinate systems for raster data.

The output is in units of square meters (m^2), showing that the result represents two-dimensional space. This information, stored as an attribute (which interested readers can discover with `attributes(st_area(luxembourg))`), can feed into subsequent calculations that use units, such as population density (which is measured in people per unit area, typically per km^2). Reporting units prevents confusion. To take the Luxembourg example, if the units remained unspecified, one could incorrectly assume that the units were in hectares. To translate the huge number into a more digestible size, it is tempting to divide the results by a million (the number of square meters in a square kilometer):

```
st_area(luxembourg) / 1000000
#> 2417 [m^2]
```

However, the result is incorrectly given again as square meters. The solution is to set the correct units with the **units** package:

```
units::set_units(st_area(luxembourg), km^2)
#> 2417 [km^2]
```

Units are of equal importance in the case of raster data. However, so far **sf** is the only spatial package that supports units, meaning that people working on raster data should approach changes in the units of analysis (for example, converting pixel widths from imperial to decimal units) with care. The `new_raster` object (see above) uses a WGS84 projection with decimal degrees as units.

Consequently, its resolution is also given in decimal degrees but you have to know it, since the `res()` function simply returns a numeric vector.

```
res(new_raster)
#> [1] 0.000833 0.000833
```

If we used the UTM projection, the units would change.

```
repr = projectRaster(new_raster, crs = "+init=epsg:26912")
res(repr)
#> [1] 0.000833 0.000833
```

Again, the `res()` command gives back a numeric vector without any unit, forcing us to know that the unit of the UTM projection is meters.

2.6 Exercises

1. Use `summary()` on the geometry column of the `world` data object. What does the output tell us about:
   - Its geometry type?
   - The number of countries?
   - Its coordinate reference system (CRS)?
2. Run the code that 'generated' the map of the world in Figure 2.5 at the end of Section 2.2.4. Find two similarities and two differences between the image on your computer and that in the book.
   - What does the `cex` argument do (see `?plot`)?
   - Why was `cex` set to the `sqrt(world$pop) / 10000`?
   - Bonus: experiment with different ways to visualize the global population.
3. Use `plot()` to create maps of Nigeria in context (see Section 2.2.4).
   - Adjust the `lwd`, `col` and `expandBB` arguments of `plot()`.
   - Challenge: read the documentation of `text()` and annotate the map.
4. Create an empty `RasterLayer` object called `my_raster` with 10 columns and 10 rows. Assign random values between 0 and 10 to the new raster and plot it.
5. Read-in the `raster/nlcd2011.tif` file from the **spDataLarge** package. What kind of information can you get about the properties of this file?

Reminder: solutions can be found online at `https://geocompr.github.io`

3

Attribute data operations

Prerequisites

- This chapter requires the following packages to be installed and attached:

```
library(sf)
library(raster)
library(dplyr)
library(stringr) # for working with strings (pattern matching)
```

- It also relies on **spData**, which loads datasets used in the code examples of this chapter:

```
library(spData)
```

3.1 Introduction

Attribute data is non-spatial information associated with geographic (geometry) data. A bus stop provides a simple example: its position would typically be represented by latitude and longitude coordinates (geometry data), in addition to its name. The name is an *attribute* of the feature (to use Simple Features terminology) that bears no relation to its geometry.

Another example is the elevation value (attribute) for a specific grid cell in raster data. Unlike the vector data model, the raster data model stores the coordinate of the grid cell indirectly, meaning the distinction between attribute and spatial information is less clear. To illustrate the point, think of a pixel in the 3rd row and the 4th column of a raster matrix. Its spatial location is defined by its index in the matrix: move from the origin four cells in the x direction (typically east and right on maps) and three cells in the y direction (typically south and down). The raster's *resolution* defines the distance for each x- and y-step which is specified in a *header*. The header is a vital component of raster datasets which specifies how pixels relate to geographic coordinates (see also Chapter 4).

The focus of this chapter is manipulating geographic objects based on attributes such as the name of a bus stop and elevation. For vector data, this means operations such as subsetting and aggregation (see Sections 3.2.1 and 3.2.2). These non-spatial operations have spatial equivalents: the [operator in base R, for example, works equally for subsetting objects based on their attribute and spatial objects, as we will see in Chapter 4. This is good news: skills developed here are cross-transferable, meaning that this chapter lays the foundation for Chapter 4, which extends the methods presented here to the spatial world. Sections 3.2.3 and 3.2.4 demonstrate how to join data onto simple feature objects using a shared ID and how to create new variables, respectively.

Raster attribute data operations are covered in Section 3.3, which covers creating continuous and categorical raster layers and extracting cell values from one layer and multiple layers (raster subsetting). Section 3.3.2 provides an overview of 'global' raster operations which can be used to characterize entire raster datasets.

3.2   Vector attribute manipulation

Geographic vector data in R are well supported by `sf`, a class which extends the `data.frame`. Thus `sf` objects have one column per attribute variable (such as 'name') and one row per observation, or *feature* (e.g., per bus station). `sf` objects also have a special column to contain geometry data, usually named `geometry`. The `geometry` column is special because it is a *list column*, which can contain multiple geographic entities (points, lines, polygons) per row. This was described in Chapter 2, which demonstrated how *generic methods* such as `plot()` and `summary()` work on `sf` objects. **sf** also provides methods that allow `sf` objects to behave like regular data frames, as illustrated by other `sf`-specific methods that were originally developed for data frames:

```
methods(class = "sf") # methods for sf objects, first 12 shown
```

```
#>   [1] aggregate        cbind            coerce
#>   [4] initialize       merge            plot
#>   [7] print            rbind            [
#>  [10] [[<-             $<-              show
```

Many of these functions, including `rbind()` (for binding rows of data together) and `$<-` (for creating new columns) were developed for data frames. A key feature of sf objects is that they store spatial and non-spatial data in the same way, as columns in a data.frame.

> The geometry column of sf objects is typically called geometry but any name can be used. The following command, for example, creates a geometry column named g:
>
> ```
> st_sf(data.frame(n = world$name_long), g = world$geom)
> ```
>
> This enables geometries imported from spatial databases to have a variety of names such as wkb_geometry and the_geom.

sf objects also support tibble and tbl classes used in the tidyverse, allowing 'tidy' data analysis workflows for spatial data. Thus **sf** enables the full power of R's data analysis capabilities to be unleashed on geographic data. Before using these capabilities it is worth re-capping how to discover the basic properties of vector data objects. Let's start by using base R functions to get a measure of the world dataset:

```
dim(world) # it is a 2 dimensional object, with rows and columns
#> [1] 177   11
nrow(world) # how many rows?
#> [1] 177
ncol(world) # how many columns?
#> [1] 11
```

Our dataset contains ten non-geographic columns (and one geometry list column) with almost 200 rows representing the world's countries. Extracting the attribute data of an sf object is the same as removing its geometry:

```
world_df = st_drop_geometry(world)
class(world_df)
#> [1] "data.frame"
```

This can be useful if the geometry column causes problems, e.g., by occupying large amounts of RAM, or to focus the attention on the attribute data. For most cases, however, there is no harm in keeping the geometry column because non-spatial data operations on sf objects only change an object's geometry when appropriate (e.g., by dissolving borders between adjacent polygons following aggregation). This means that proficiency with attribute data in sf objects equates to proficiency with data frames in R.

For many applications, the tidyverse package **dplyr** offers the most effective and intuitive approach for working with data frames. Tidyverse compatibility is an advantage of **sf** over its predecessor **sp**, but there are some pitfalls to avoid (see the supplementary `tidyverse-pitfalls` vignette at geocompr.github.io[1] for details).

3.2.1 Vector attribute subsetting

Base R subsetting functions include `[`, `subset()` and `$`. **dplyr** subsetting functions include `select()`, `filter()`, and `pull()`. Both sets of functions preserve the spatial components of attribute data in `sf` objects.

The `[` operator can subset both rows and columns. You use indices to specify the elements you wish to extract from an object, e.g., `object[i, j]`, with `i` and `j` typically being numbers or logical vectors — `TRUE`s and `FALSE`s — representing rows and columns (they can also be character strings, indicating row or column names). Leaving `i` or `j` empty returns all rows or columns, so `world[1:5, ]` returns the first five rows and all columns. The examples below demonstrate subsetting with base R. The results are not shown; check the results on your own computer:

```
world[1:6, ] # subset rows by position
world[, 1:3] # subset columns by position
world[, c("name_long", "lifeExp")] # subset columns by name
```

A demonstration of the utility of using `logical` vectors for subsetting is shown in the code chunk below. This creates a new object, `small_countries`, containing nations whose surface area is smaller than 10,000 km^2:

```
sel_area = world$area_km2 < 10000
summary(sel_area) # a logical vector
#>    Mode   FALSE    TRUE
#> logical    170       7
small_countries = world[sel_area, ]
```

The intermediary `sel_area` is a logical vector that shows that only seven countries match the query. A more concise command, which omits the intermediary object, generates the same result:

```
small_countries = world[world$area_km2 < 10000, ]
```

The base R function `subset()` provides yet another way to achieve the same result:

---

[1] https://geocompr.github.io/geocompkg/articles/tidyverse-pitfalls.html

```
small_countries = subset(world, area_km2 < 10000)
```

Base R functions are mature and widely used. However, the more recent **dplyr** approach has several advantages. It enables intuitive workflows. It is fast, due to its C++ backend. This is especially useful when working with big data as well as **dplyr**'s database integration. The main **dplyr** subsetting functions are `select()`, `slice()`, `filter()` and `pull()`.

raster and **dplyr** packages have a function called `select()`. When using both packages, the function in the most recently attached package will be used, 'masking' the incumbent function. This can generate error messages containing text like: `unable to find an inherited method for function 'select' for signature '"sf"'`. To avoid this error message, and prevent ambiguity, we use the long-form function name, prefixed by the package name and two colons (usually omitted from R scripts for concise code): `dplyr::select()`.

`select()` selects columns by name or position. For example, you could select only two columns, `name_long` and `pop`, with the following command (note the sticky `geom` column remains):

```
world1 = dplyr::select(world, name_long, pop)
names(world1)
#> [1] "name_long" "pop"        "geom"
```

`select()` also allows subsetting of a range of columns with the help of the : operator:

```
# all columns between name_long and pop (inclusive)
world2 = dplyr::select(world, name_long:pop)
```

Omit specific columns with the - operator:

```
# all columns except subregion and area_km2 (inclusive)
world3 = dplyr::select(world, -subregion, -area_km2)
```

Conveniently, `select()` lets you subset and rename columns at the same time, for example:

```
world4 = dplyr::select(world, name_long, population = pop)
names(world4)
#> [1] "name_long"  "population" "geom"
```

This is more concise than the base R equivalent:

```
world5 = world[, c("name_long", "pop")] # subset columns by name
names(world5)[names(world5) == "pop"] = "population" # rename column manually
```

`select()` also works with 'helper functions' for advanced subsetting operations, including `contains()`, `starts_with()` and `num_range()` (see the help page with `?select` for details).

Most **dplyr** verbs return a data frame. To extract a single vector, one has to explicitly use the `pull()` command. The subsetting operator in base R (see `?[`), by contrast, tries to return objects in the lowest possible dimension. This means selecting a single column returns a vector in base R. To turn off this behavior, set the `drop` argument to FALSE.

```
# create throw-away data frame
d = data.frame(pop = 1:10, area = 1:10)
# return data frame object when selecting a single column
d[, "pop", drop = FALSE] # equivalent to d["pop"]
select(d, pop)
# return a vector when selecting a single column
d[, "pop"]
pull(d, pop)
```

Due to the sticky geometry column, selecting a single attribute from an sf-object with the help of `[()` returns also a data frame. Contrastingly, `pull()` and `$` will give back a vector.

```
# data frame object
world[, "pop"]
# vector objects
world$pop
pull(world, pop)
```

`slice()` is the row-equivalent of `select()`. The following code chunk, for example, selects the 3rd to 5th rows:

```
slice(world, 3:5)
```

`filter()` is **dplyr**'s equivalent of base R's `subset()` function. It keeps only rows matching given criteria, e.g., only countries with a very high average of life expectancy:

TABLE 3.1: Comparison operators that return Booleans (TRUE/FALSE).

Symbol	Name
'=='	Equal to
'!='	Not equal to
'>', '<'	Greater/Less than
'>=', '<='	Greater/Less than or equal
'&', '|', '!'	Logical operators: And, Or, Not

```
# Countries with a life expectancy longer than 82 years
world6 = filter(world, lifeExp > 82)
```

The standard set of comparison operators can be used in the `filter()` function, as illustrated in Table 3.1:

dplyr works well with the 'pipe'[2] operator `%>%`, which takes its name from the Unix pipe | (Grolemund and Wickham, 2016). It enables expressive code: the output of a previous function becomes the first argument of the next function, enabling *chaining*. This is illustrated below, in which only countries from Asia are filtered from the `world` dataset, next the object is subset by columns (`name_long` and `continent`) and the first five rows (result not shown).

```
world7 = world %>%
  filter(continent == "Asia") %>%
  dplyr::select(name_long, continent) %>%
  slice(1:5)
```

The above chunk shows how the pipe operator allows commands to be written in a clear order: the above run from top to bottom (line-by-line) and left to right. The alternative to `%>%` is nested function calls, which is harder to read:

```
world8 = slice(
  dplyr::select(
    filter(world, continent == "Asia"),
    name_long, continent),
  1:5)
```

---

[2]http://r4ds.had.co.nz/pipes.html

3.2.2   Vector attribute aggregation

Aggregation operations summarize datasets by a 'grouping variable', typically an attribute column (spatial aggregation is covered in the next chapter). An example of attribute aggregation is calculating the number of people per continent based on country-level data (one row per country). The `world` dataset contains the necessary ingredients: the columns `pop` and `continent`, the population and the grouping variable, respectively. The aim is to find the `sum()` of country populations for each continent. This can be done with the base R function `aggregate()` as follows:

```
world_agg1 = aggregate(pop ~ continent, FUN = sum, data = world, na.rm = TRUE)
class(world_agg1)
#> [1] "data.frame"
```

The result is a non-spatial data frame with six rows, one per continent, and two columns reporting the name and population of each continent (see Table 3.2 with results for the top 3 most populous continents).

`aggregate()` is a generic function which means that it behaves differently depending on its inputs. **sf** provides a function that can be called directly with `sf:::aggregate()` that is activated when a `by` argument is provided, rather than using the `~` to refer to the grouping variable:

```
world_agg2 = aggregate(world["pop"], by = list(world$continent),
                       FUN = sum, na.rm = TRUE)
class(world_agg2)
#> [1] "sf"          "data.frame"
```

As illustrated above, an object of class `sf` is returned this time. `world_agg2` which is a spatial object containing 6 polygons representing the columns of the world.

`summarize()` is the **dplyr** equivalent of `aggregate()`. It usually follows `group_by()`, which specifies the grouping variable, as illustrated below:

```
world_agg3 = world %>%
  group_by(continent) %>%
  summarize(pop = sum(pop, na.rm = TRUE))
```

This approach is flexible and gives control over the new column names. This is illustrated below: the command calculates the Earth's population (~7 billion) and number of countries (result not shown):

TABLE 3.2: The top 3 most populous continents, and the number of countries in each.

continent	pop	n_countries
Africa	1154946633	51
Asia	4311408059	47
Europe	669036256	39

```
world %>%
  summarize(pop = sum(pop, na.rm = TRUE), n = n())
```

In the previous code chunk `pop` and `n` are column names in the result. `sum()` and `n()` were the aggregating functions. The result is an `sf` object with a single row representing the world (this works thanks to the geometric operation 'union', as explained in Section 5.2.6).

Let's combine what we have learned so far about **dplyr** by chaining together functions to find the world's 3 most populous continents (with `dplyr::top_n()`) and the number of countries they contain (the result of this command is presented in Table 3.2):

```
world %>%
  dplyr::select(pop, continent) %>%
  group_by(continent) %>%
  summarize(pop = sum(pop, na.rm = TRUE), n_countries = n()) %>%
  top_n(n = 3, wt = pop) %>%
  st_drop_geometry()
```

More details are provided in the help pages (which can be accessed via `?summarize` and `vignette(package = "dplyr")`) and Chapter 5 of R for Data Science[3].

3.2.3 Vector attribute joining

Combining data from different sources is a common task in data preparation. Joins do this by combining tables based on a shared 'key' variable. **dplyr** has multiple join functions including `left_join()` and `inner_join()` — see `vignette("two-table")` for a full list. These function names follow conventions used in the database language SQL[4] (Grolemund and Wickham, 2016, Chapter 13); using them to join non-spatial datasets to `sf` objects is the focus of this section.

---

[4]http://r4ds.had.co.nz/relational-data.html

dplyr join functions work the same on data frames and `sf` objects, the only important difference being the `geometry` list column. The result of data joins can be either an `sf` or `data.frame` object. The most common type of attribute join on spatial data takes an `sf` object as the first argument and adds columns to it from a `data.frame` specified as the second argument.

To demonstrate joins, we will combine data on coffee production with the `world` dataset. The coffee data is in a data frame called `coffee_data` from the **spData** package (see `?coffee_data` for details). It has 3 columns: `name_long` names major coffee-producing nations and `coffee_production_2016` and `coffee_production_2017` contain estimated values for coffee production in units of 60-kg bags in each year. A 'left join', which preserves the first dataset, merges `world` with `coffee_data`:

```
world_coffee = left_join(world, coffee_data)
#> Joining, by = "name_long"
class(world_coffee)
#> [1] "sf"          "data.frame"
```

Because the input datasets share a 'key variable' (`name_long`) the join worked without using the `by` argument (see `?left_join` for details). The result is an `sf` object identical to the original `world` object but with two new variables (with column indices 11 and 12) on coffee production. This can be plotted as a map, as illustrated in Figure 3.1, generated with the `plot()` function below:

```
names(world_coffee)
#>  [1] "iso_a2"                 "name_long"
#>  [3] "continent"              "region_un"
#>  [5] "subregion"              "type"
#>  [7] "area_km2"               "pop"
#>  [9] "lifeExp"                "gdpPercap"
#> [11] "coffee_production_2016" "coffee_production_2017"
#> [13] "geom"
plot(world_coffee["coffee_production_2017"])
```

For joining to work, a 'key variable' must be supplied in both datasets. By default **dplyr** uses all variables with matching names. In this case, both `world_coffee` and `world` objects contained a variable called `name_long`, explaining the message `Joining, by = "name_long"`. In the majority of cases where variable names are not the same, you have two options:

1. Rename the key variable in one of the objects so they match.
2. Use the `by` argument to specify the joining variables.

coffee_production_2017

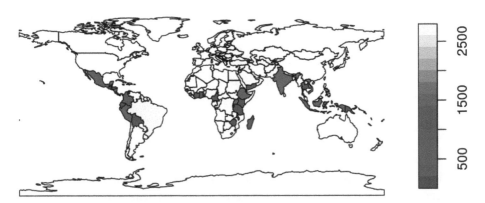

FIGURE 3.1: World coffee production (thousand 60-kg bags) by country, 2017. Source: International Coffee Organization.

The latter approach is demonstrated below on a renamed version of `coffee_data`:

```
coffee_renamed = rename(coffee_data, nm = name_long)
world_coffee2 = left_join(world, coffee_renamed, by = c(name_long = "nm"))
```

Note that the name in the original object is kept, meaning that `world_coffee` and the new object `world_coffee2` are identical. Another feature of the result is that it has the same number of rows as the original dataset. Although there are only 47 rows of data in `coffee_data`, all 177 the country records are kept intact in `world_coffee` and `world_coffee2`: rows in the original dataset with no match are assigned `NA` values for the new coffee production variables. What if we only want to keep countries that have a match in the key variable? In that case an inner join can be used:

```
world_coffee_inner = inner_join(world, coffee_data)
#> Joining, by = "name_long"
nrow(world_coffee_inner)
#> [1] 45
```

Note that the result of `inner_join()` has only 45 rows compared with 47 in `coffee_data`. What happened to the remaining rows? We can identify the rows that did not match using the `setdiff()` function as follows:

```
setdiff(coffee_data$name_long, world$name_long)
#> [1] "Congo, Dem. Rep. of" "Others"
```

The result shows that `Others` accounts for one row not present in the `world` dataset and that the name of the `Democratic Republic of the Congo` accounts for the other: it has been abbreviated, causing the join to miss it. The following command uses a string matching (regex) function from the **stringr** package to confirm what `Congo, Dem. Rep. of` should be:

```
str_subset(world$name_long, "Dem*.+Congo")
#> [1] "Democratic Republic of the Congo"
```

To fix this issue, we will create a new version of `coffee_data` and update the name. `inner_join()`ing the updated data frame returns a result with all 46 coffee-producing nations:

```
coffee_data$name_long[grepl("Congo,", coffee_data$name_long)] =
  str_subset(world$name_long, "Dem*.+Congo")
world_coffee_match = inner_join(world, coffee_data)
#> Joining, by = "name_long"
nrow(world_coffee_match)
#> [1] 46
```

It is also possible to join in the other direction: starting with a non-spatial dataset and adding variables from a simple features object. This is demonstrated below, which starts with the `coffee_data` object and adds variables from the original `world` dataset. In contrast with the previous joins, the result is *not* another simple feature object, but a data frame in the form of a **tidyverse** tibble: the output of a join tends to match its first argument:

```
coffee_world = left_join(coffee_data, world)
#> Joining, by = "name_long"
class(coffee_world)
#> [1] "tbl_df"    "tbl"       "data.frame"
```

 In most cases, the geometry column is only useful in an `sf` object. The geometry column can only be used for creating maps and spatial operations if R 'knows' it is a spatial object, defined by a spatial package such as **sf**. Fortunately, non-spatial data frames with a geometry list column (like `coffee_world`) can be coerced into an `sf` object as follows: `st_as_sf(coffee_world)`.

This section covers the majority joining use cases. For more information, we recommend Grolemund and Wickham (2016), the join vignette[5] in the **geocompkg** package that accompanies this book, and documentation of the **data.table** package.[6] Another type of join is a spatial join, covered in the next chapter (Section 4.2.3).

3.2.4 Creating attributes and removing spatial information

Often, we would like to create a new column based on already existing columns. For example, we want to calculate population density for each country. For this we need to divide a population column, here pop, by an area column, here area_km2 with unit area in square kilometers. Using base R, we can type:

```
world_new = world # do not overwrite our original data
world_new$pop_dens = world_new$pop / world_new$area_km2
```

Alternatively, we can use one of **dplyr** functions - mutate() or transmute(). mutate() adds new columns at the penultimate position in the sf object (the last one is reserved for the geometry):

```
world %>%
  mutate(pop_dens = pop / area_km2)
```

The difference between mutate() and transmute() is that the latter skips all other existing columns (except for the sticky geometry column):

```
world %>%
  transmute(pop_dens = pop / area_km2)
```

unite() pastes together existing columns. For example, we want to combine the continent and region_un columns into a new column named con_reg. Additionally, we can define a separator (here: a colon :) which defines how the values of the input columns should be joined, and if the original columns should be removed (here: TRUE):

```
world_unite = world %>%
  unite("con_reg", continent:region_un, sep = ":", remove = TRUE)
```

The separate() function does the opposite of unite(): it splits one column into multiple columns using either a regular expression or character positions.

---

[5]https://geocompr.github.io/geocompkg/articles/join.html

[6]**data.table** is a high-performance data processing package. Its application to geographic data is covered in a blog post hosted at r-spatial.org/r/2017/11/13/perp-performance.html.

```
world_separate = world_unite %>%
  separate(con_reg, c("continent", "region_un"), sep = ":")
```

The two functions `rename()` and `set_names()` are useful for renaming columns.
The first replaces an old name with a new one. The following command, for
example, renames the lengthy `name_long` column to simply `name`:

```
world %>%
  rename(name = name_long)
```

`set_names()` changes all column names at once, and requires a character vector
with a name matching each column. This is illustrated below, which outputs
the same `world` object, but with very short names:

```
new_names = c("i", "n", "c", "r", "s", "t", "a", "p", "l", "gP", "geom")
world %>%
  set_names(new_names)
```

It is important to note that attribute data operations preserve the geometry
of the simple features. As mentioned at the outset of the chapter, it can be
useful to remove the geometry. To do this, you have to explicitly remove it
because `sf` explicitly makes the geometry column sticky. This behavior ensures
that data frame operations do not accidentally remove the geometry column.
Hence, an approach such as `select(world, -geom)` will be unsuccessful and you
should instead use `st_drop_geometry()`.[7]

```
world_data = world %>% st_drop_geometry()
class(world_data)
#> [1] "data.frame"
```

3.3 Manipulating raster objects

In contrast to the vector data model underlying simple features (which represents points, lines and polygons as discrete entities in space), raster data
represent continuous surfaces. This section shows how raster objects work by
creating them *from scratch*, building on Section 2.3.1. Because of their unique
structure, subsetting and other operations on raster datasets work in a different
way, as demonstrated in Section 3.3.1.

---

[7] `st_geometry(world_st) = NULL` also works to remove the geometry from `world`, but overwrites
the original object.

The following code recreates the raster dataset used in Section 2.3.3, the result of which is illustrated in Figure 3.2. This demonstrates how the `raster()` function works to create an example raster named `elev` (representing elevations).

```
elev = raster(nrows = 6, ncols = 6, res = 0.5,
              xmn = -1.5, xmx = 1.5, ymn = -1.5, ymx = 1.5,
              vals = 1:36)
```

The result is a raster object with 6 rows and 6 columns (specified by the `nrow` and `ncol` arguments), and a minimum and maximum spatial extent in x and y direction (`xmn`, `xmx`, `ymn`, `ymax`). The `vals` argument sets the values that each cell contains: numeric data ranging from 1 to 36 in this case. Raster objects can also contain categorical values of class `logical` or `factor` variables in R. The following code creates a raster representing grain sizes (Figure 3.2):

```
grain_order = c("clay", "silt", "sand")
grain_char = sample(grain_order, 36, replace = TRUE)
grain_fact = factor(grain_char, levels = grain_order)
grain = raster(nrows = 6, ncols = 6, res = 0.5,
               xmn = -1.5, xmx = 1.5, ymn = -1.5, ymx = 1.5,
               vals = grain_fact)
```

`raster` objects can contain values of class `numeric`, `integer`, `logical` or `factor`, but not `character`. To use character values, they must first be converted into an appropriate class, for example using the function `factor()`. The `levels` argument was used in the preceding code chunk to create an ordered factor: clay < silt < sand in terms of grain size. See the Data structures chapter of Wickham (2014a) for further details on classes.

`raster` objects represent categorical variables as integers, so `grain[1, 1]` returns a number that represents a unique identifier, rather than "clay", "silt" or "sand". The raster object stores the corresponding look-up table or "Raster Attribute Table" (RAT) as a data frame in a new slot named `attributes`, which can be viewed with `ratify(grain)` (see `?ratify()` for more information). Use the function `levels()` for retrieving and adding new factor levels to the attribute table:

```
levels(grain)[[1]] = cbind(levels(grain)[[1]], wetness = c("wet", "moist", "dry"))
levels(grain)
#> [[1]]
#>    ID VALUE wetness
#> 1  1  clay     wet
```

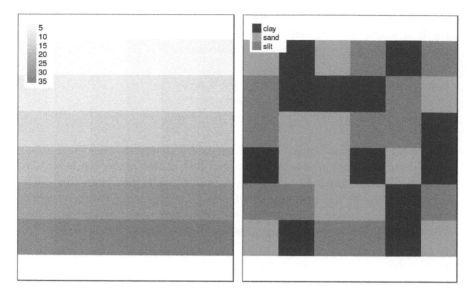

FIGURE 3.2: Raster datasets with numeric (left) and categorical values
(right).

```
#> 2  2  silt    moist
#> 3  3  sand    dry
```

This behavior demonstrates that raster cells can only possess one value, an
identifier which can be used to look up the attributes in the corresponding
attribute table (stored in a slot named attributes). This is illustrated in
command below, which returns the grain size and wetness of cell IDs 1, 11 and
35:

```
factorValues(grain, grain[c(1, 11, 35)])
#>    VALUE wetness
#> 1  sand     dry
#> 2  silt   moist
#> 3  clay     wet
```

3.3.1   Raster subsetting

Raster subsetting is done with the base R operator [, which accepts a variety
of inputs:

- Row-column indexing
- Cell IDs

- Coordinates
- Another raster object

Here, we only show the first two options since these can be considered non-spatial operations. If we need a spatial object to subset another or the output is a spatial object, we refer to this as spatial subsetting. Therefore, the latter two options will be shown in the next chapter (see Section 4.3.1 in the next chapter).

The first two subsetting options are demonstrated in the commands below — both return the value of the top left pixel in the raster object `elev` (results not shown):

```
# row 1, column 1
elev[1, 1]
# cell ID 1
elev[1]
```

To extract all values or complete rows, you can use `values()` and `getValues()`. For multi-layered raster objects `stack` or `brick`, this will return the cell value(s) for each layer. For example, `stack(elev, grain)[1]` returns a matrix with one row and two columns — one for each layer. For multi-layer raster objects another way to subset is with `raster::subset()`, which extracts layers from a raster stack or brick. The `[[` and `$` operators can also be used:

```
r_stack = stack(elev, grain)
names(r_stack) = c("elev", "grain")
# three ways to extract a layer of a stack
raster::subset(r_stack, "elev")
r_stack[["elev"]]
r_stack$elev
```

Cell values can be modified by overwriting existing values in conjunction with a subsetting operation. The following code chunk, for example, sets the upper left cell of `elev` to 0:

```
elev[1, 1] = 0
elev[]
#>  [1]  0  2  3  4  5  6  7  8  9 10 11 12 13 14 15 16 17 18 19 20 21 22 23
#> [24] 24 25 26 27 28 29 30 31 32 33 34 35 36
```

Leaving the square brackets empty is a shortcut version of `values()` for retrieving all values of a raster. Multiple cells can also be modified in this way:

```
elev[1, 1:2] = 0
```

3.3.2   Summarizing raster objects

raster contains functions for extracting descriptive statistics for entire rasters. Printing a raster object to the console by typing its name returns minimum and maximum values of a raster. `summary()` provides common descriptive statistics (minimum, maximum, interquartile range and number of NAs). Further summary operations such as the standard deviation (see below) or custom summary statistics can be calculated with `cellStats()`.

```
cellStats(elev, sd)
```

> If you provide the `summary()` and `cellStats()` functions with a raster stack or brick object, they will summarize each layer separately, as can be illustrated by running: `summary(brick(elev, grain))`.

Raster value statistics can be visualized in a variety of ways. Specific functions such as `boxplot()`, `density()`, `hist()` and `pairs()` work also with raster objects, as demonstrated in the histogram created with the command below (not shown):

```
hist(elev)
```

In case a visualization function does not work with raster objects, one can extract the raster data to be plotted with the help of `values()` or `getValues()`.

Descriptive raster statistics belong to the so-called global raster operations. These and other typical raster processing operations are part of the map algebra scheme, which are covered in the next chapter (Section 4.3.2).

> Some function names clash between packages (e.g., `select()`, as discussed in a previous note). In addition to not loading packages by referring to functions verbosely (e.g., `dplyr::select()`), another way to prevent function names clashes is by unloading the offending package with `detach()`. The following command, for example, unloads the **raster** package (this can also be done in the *package* tab which resides by default in the right-bottom pane in RStudio): `detach("package:raster", unload = TRUE, force = TRUE)`. The `force` argument makes sure that the package will be detached even if other packages depend on it. This, however, may lead to a restricted usability of packages depending on the detached package, and is therefore not recommended.

3.4 Exercises

For these exercises we will use the `us_states` and `us_states_df` datasets from the **spData** package:

```
library(spData)
data(us_states)
data(us_states_df)
```

`us_states` is a spatial object (of class `sf`), containing geometry and a few attributes (including name, region, area, and population) of states within the contiguous United States. `us_states_df` is a data frame (of class `data.frame`) containing the name and additional variables (including median income and poverty level, for the years 2010 and 2015) of US states, including Alaska, Hawaii and Puerto Rico. The data comes from the United States Census Bureau, and is documented in `?us_states` and `?us_states_df`.

1. Create a new object called `us_states_name` that contains only the `NAME` column from the `us_states` object. What is the class of the new object and what makes it geographic?
2. Select columns from the `us_states` object which contain population data. Obtain the same result using a different command (bonus: try to find three ways of obtaining the same result). Hint: try to use helper functions, such as `contains` or `starts_with` from **dplyr** (see `?contains`).
3. Find all states with the following characteristics (bonus find *and* plot them):
   - Belong to the Midwest region.
   - Belong to the West region, have an area below 250,000 km^2 *and* in 2015 a population greater than 5,000,000 residents (hint: you may need to use the function `units::set_units()` or `as.numeric()`).
   - Belong to the South region, had an area larger than 150,000 km^2 or a total population in 2015 larger than 7,000,000 residents.
4. What was the total population in 2015 in the `us_states` dataset? What was the minimum and maximum total population in 2015?
5. How many states are there in each region?
6. What was the minimum and maximum total population in 2015 in each region? What was the total population in 2015 in each region?
7. Add variables from `us_states_df` to `us_states`, and create a new object called `us_states_stats`. What function did you use and why? Which variable is the key in both datasets? What is the class of the new object?

8. `us_states_df` has two more rows than `us_states`. How can you find them? (hint: try to use the `dplyr::anti_join()` function)

9. What was the population density in 2015 in each state? What was the population density in 2010 in each state?

10. How much has population density changed between 2010 and 2015 in each state? Calculate the change in percentages and map them.

11. Change the columns' names in `us_states` to lowercase. (Hint: helper functions - `tolower()` and `colnames()` may help.)

12. Using `us_states` and `us_states_df` create a new object called `us_states_sel`. The new object should have only two variables - `median_income_15` and `geometry`. Change the name of the `median_income_15` column to `Income`.

13. Calculate the change in median income between 2010 and 2015 for each state. Bonus: What was the minimum, average and maximum median income in 2015 for each region? What is the region with the largest increase of the median income?

14. Create a raster from scratch with nine rows and columns and a resolution of 0.5 decimal degrees (WGS84). Fill it with random numbers. Extract the values of the four corner cells.

15. What is the most common class of our example raster `grain` (hint: `modal()`)?

16. Plot the histogram and the boxplot of the `data(dem, package = "RQGIS")` raster.

4

Spatial data operations

Prerequisites

- This chapter requires the same packages used in Chapter 3:

```
library(sf)
library(raster)
library(dplyr)
library(spData)
```

4.1 Introduction

Spatial operations are a vital part of geocomputation. This chapter shows how spatial objects can be modified in a multitude of ways based on their location and shape. The content builds on the previous chapter because many spatial operations have a non-spatial (attribute) equivalent. This is especially true for *vector* operations: Section 3.2 on vector attribute manipulation provides the basis for understanding its spatial counterpart, namely spatial subsetting (covered in Section 4.2.1). Spatial joining (Section 4.2.3) and aggregation (Section 4.2.5) also have non-spatial counterparts, covered in the previous chapter.

Spatial operations differ from non-spatial operations in some ways, however. To illustrate the point, imagine you are researching road safety. Spatial joins can be used to find road speed limits related with administrative zones, even when no zone ID is provided. But this raises the question: should the road completely fall inside a zone for its values to be joined? Or is simply crossing or being within a certain distance sufficient? When posing such questions, it becomes apparent that spatial operations differ substantially from attribute operations on data frames: the *type* of spatial relationship between objects must be considered. These are covered in Section 4.2.2, on topological relations.

Another unique aspect of spatial objects is distance. All spatial objects are related through space and distance calculations, covered in Section 4.2.6, can be used to explore the strength of this relationship.

Spatial operations also apply to raster objects. Spatial subsetting of raster objects is covered in Section 4.3.1; merging several raster 'tiles' into a single object is covered in Section 4.3.7. For many applications, the most important spatial operation on raster objects is *map algebra*, as we will see in Sections 4.3.2 to 4.3.6. Map algebra is also the prerequisite for distance calculations on rasters, a technique which is covered in Section 4.3.6.

It is important to note that spatial operations that use two spatial objects rely on both objects having the same coordinate reference system, a topic that was introduced in Section 2.4 and which will be covered in more depth in Chapter 6.

4.2 Spatial operations on vector data

This section provides an overview of spatial operations on vector geographic data represented as simple features in the **sf** package before Section 4.3, which presents spatial methods using the **raster** package.

4.2.1 Spatial subsetting

Spatial subsetting is the process of selecting features of a spatial object based on whether or not they in some way *relate* in space to another object. It is analogous to *attribute subsetting* (covered in Section 3.2.1) and can be done with the base R square bracket (`[`) operator or with the `filter()` function from the **tidyverse**.

An example of spatial subsetting is provided by the `nz` and `nz_height` datasets in **spData**. These contain projected data on the 16 main regions and 101 highest points in New Zealand, respectively (Figure 4.1). The following code chunk first creates an object representing Canterbury, then uses spatial subsetting to return all high points in the region:

```
canterbury = nz %>% filter(Name == "Canterbury")
canterbury_height = nz_height[canterbury, ]
```

Like attribute subsetting `x[y, ]` subsets features of a *target* `x` using the contents of a *source* object `y`. Instead of `y` being of class `logical` or `integer` — a vector of

High points in New Zealand

High points in Canterbury

FIGURE 4.1: Illustration of spatial subsetting with red triangles representing 101 high points in New Zealand, clustered near the central Canterbuy region (left). The points in Canterbury were created with the '[' subsetting operator (highlighted in gray, right).

TRUE and FALSE values or whole numbers — for spatial subsetting it is another spatial (sf) object.

Various *topological relations* can be used for spatial subsetting. These determine the type of spatial relationship that features in the target object must have with the subsetting object to be selected, including *touches*, *crosses* or *within* (see Section 4.2.2). *Intersects* is the default spatial subsetting operator, a default that returns TRUE for many types of spatial relations, including *touches*, *crosses* and *is within*. These alternative spatial operators can be specified with the op = argument, a third argument that can be passed to the [operator for sf objects. This is demonstrated in the following command which returns the opposite of st_intersect(), points that do not intersect with Canterbury (see Section 4.2.2):

```
nz_height[canterbury, , op = st_disjoint]
```

Note the empty argument — denoted with , , — in the preceding code chunk is included to highlight op, the third argument in [for sf objects. One can use this to change the subsetting operation in many ways. nz_height[canterbury,

`2, op = st_disjoint]`, for example, returns the same rows but only includes the second attribute column (see `sf:::'[.sf'` and the `?sf` for details).

For many applications, this is all you'll need to know about spatial subsetting for vector data. In this case, you can safely skip to Section 4.2.2.

If you're interested in the details, including other ways of subsetting, read on. Another way of doing spatial subsetting uses objects returned by *topological operators*. This is demonstrated in the first command below:

```
sel_sgbp = st_intersects(x = nz_height, y = canterbury)
class(sel_sgbp)
#> [1] "sgbp"
sel_logical = lengths(sel_sgbp) > 0
canterbury_height2 = nz_height[sel_logical, ]
```

In the above code chunk, an object of class `sgbp` (a sparse geometry binary predicate, a list of length x in the spatial operation) is created and then converted into a logical vector `sel_logical` (containing only TRUE and FALSE values). The function `lengths()` identifies which features in `nz_height` intersect with *any* objects in y. In this case 1 is the greatest possible value but for more complex operations one could use the method to subset only features that intersect with, for example, 2 or more features from the source object.

 Note: another way to return a logical output is by setting `sparse = FALSE` (meaning 'return a dense matrix not a sparse one') in operators such as `st_intersects()`. The command `st_intersects(x = nz_height, y = canterbury, sparse = FALSE)[, 1]`, for example, would return an output identical `sel_logical`. Note: the solution involving `sgbp` objects is more generalisable though, as it works for many-to-many operations and has lower memory requirements.

It should be noted that a logical can also be used with `filter()` as follows (`sparse = FALSE` is explained in Section 4.2.2):

```
canterbury_height3 = nz_height %>%
  filter(st_intersects(x = ., y = canterbury, sparse = FALSE))
```

At this point, there are three versions of `canterbury_height`, one created with spatial subsetting directly and the other two via intermediary selection objects. To explore these objects and spatial subsetting in more detail, see the supplementary vignettes on `subsetting` and `tidverse-pitfalls`[1].

---

[1] `https://geocompr.github.io/geocompkg/articles/`

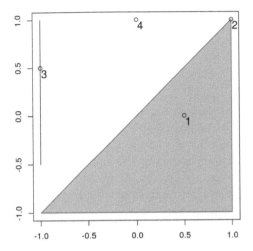

FIGURE 4.2: Points (p 1 to 4), line and polygon objects arranged to illustrate topological relations.

4.2.2 Topological relations

Topological relations describe the spatial relationships between objects. To understand them, it helps to have some simple test data to work with. Figure 4.2 contains a polygon (a), a line (l) and some points (p), which are created in the code below.

```
# create a polygon
a_poly = st_polygon(list(rbind(c(-1, -1), c(1, -1), c(1, 1), c(-1, -1))))
a = st_sfc(a_poly)
# create a line
l_line = st_linestring(x = matrix(c(-1, -1, -0.5, 1), ncol = 2))
l = st_sfc(l_line)
# create points
p_matrix = matrix(c(0.5, 1, -1, 0, 0, 1, 0.5, 1), ncol = 2)
p_multi = st_multipoint(x = p_matrix)
p = st_cast(st_sfc(p_multi), "POINT")
```

A simple query is: which of the points in p intersect in some way with polygon a? The question can be answered by inspection (points 1 and 2 are over or touch the triangle). It can also be answered by using a *spatial predicate* such as *do the objects intersect?* This is implemented in **sf** as follows:

```
st_intersects(p, a)
#> Sparse geometry binary ..., where the predicate was 'intersects'
#> 1: 1
```

```
#> 2: 1
#> 3: (empty)
#> 4: (empty)
```

The contents of the result should be as you expected: the function returns a
positive (1) result for the first two points, and a negative result (represented by
an empty vector) for the last two. What may be unexpected is that the result
comes in the form of a list of vectors. This *sparse matrix* output only registers
a relation if one exists, reducing the memory requirements of topological
operations on multi-feature objects. As we saw in the previous section, a *dense
matrix* consisting of TRUE or FALSE values for each combination of features can
also be returned when sparse = FALSE:

```
st_intersects(p, a, sparse = FALSE)
#>         [,1]
#> [1,]   TRUE
#> [2,]   TRUE
#> [3,] FALSE
#> [4,] FALSE
```

The output is a matrix in which each row represents a feature in the target
object and each column represents a feature in the selecting object. In this case,
only the first two features in p intersect with a and there is only one feature in
a so the result has only one column. The result can be used for subsetting as
we saw in Section 4.2.1.

Note that st_intersects() returns TRUE for the second feature in the object p
even though it just touches the polygon a: *intersects* is a 'catch-all' topological
operation which identifies many types of spatial relation.

The opposite of st_intersects() is st_disjoint(), which returns only objects that
do not spatially relate in any way to the selecting object (note [, 1] converts
the result into a vector):

```
st_disjoint(p, a, sparse = FALSE)[, 1]
#> [1] FALSE FALSE   TRUE   TRUE
```

st_within() returns TRUE only for objects that are completely within the selecting
object. This applies only to the first object, which is inside the triangular
polygon, as illustrated below:

```
st_within(p, a, sparse = FALSE)[, 1]
#> [1]  TRUE FALSE FALSE FALSE
```

Note that although the first point is *within* the triangle, it does not *touch* any part of its border. For this reason st_touches() only returns TRUE for the second point:

```
st_touches(p, a, sparse = FALSE)[, 1]
#> [1] FALSE  TRUE FALSE FALSE
```

What about features that do not touch, but *almost touch* the selection object? These can be selected using st_is_within_distance(), which has an additional dist argument. It can be used to set how close target objects need to be before they are selected. Note that although point 4 is one unit of distance from the nearest node of a (at point 2 in Figure 4.2), it is still selected when the distance is set to 0.9. This is illustrated in the code chunk below, the second line of which converts the lengthy list output into a logical object:

```
sel = st_is_within_distance(p, a, dist = 0.9) # can only return a sparse matrix
lengths(sel) > 0
#> [1]  TRUE  TRUE FALSE  TRUE
```

> Functions for calculating topological relations use spatial indices to largely speed up spatial query performance. They achieve that using the Sort-Tile-Recursive (STR) algorithm. The st_join function, mentioned in the next section, also uses the spatial indexing. You can learn more at https://www.r-spatial.org/r/2017/06/22/spatial-index.html.

4.2.3 Spatial joining

Joining two non-spatial datasets relies on a shared 'key' variable, as described in Section 3.2.3. Spatial data joining applies the same concept, but instead relies on shared areas of geographic space (it is also know as spatial overlay). As with attribute data, joining adds a new column to the target object (the argument x in joining functions), from a source object (y).

The process can be illustrated by an example. Imagine you have ten points randomly distributed across the Earth's surface. Of the points that are on land, which countries are they in? Random points to demonstrate spatial joining are created as follows:

```
set.seed(2018) # set seed for reproducibility
(bb_world = st_bbox(world)) # the world's bounds
#>    xmin   ymin   xmax   ymax
#> -180.0  -90.0  180.0   83.6
random_df = tibble(
```

```
  x = runif(n = 10, min = bb_world[1], max = bb_world[3]),
  y = runif(n = 10, min = bb_world[2], max = bb_world[4])
)
random_points = random_df %>%
  st_as_sf(coords = c("x", "y")) %>% # set coordinates
  st_set_crs(4326) # set geographic CRS
```

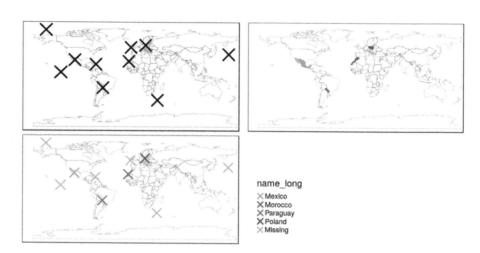

FIGURE 4.3: Illustration of a spatial join. A new attribute variable is added to random points (top left) from source world object (top right) resulting in the data represented in the final panel.

The scenario is illustrated in Figure 4.3. The `random_points` object (top left) has no attribute data, while the `world` (top right) does. The spatial join operation is done by `st_join()`, which adds the `name_long` variable to the points, resulting in `random_joined` which is illustrated in Figure 4.3 (bottom left — see `04-spatial-join.R`[2]). Before creating the joined dataset, we use spatial subsetting to create `world_random`, which contains only countries that contain random points, to verify the number of country names returned in the joined dataset should be four (see the top right panel of Figure 4.3).

```
world_random = world[random_points, ]
nrow(world_random)
#> [1] 4
random_joined = st_join(random_points, world["name_long"])
```

By default, `st_join()` performs a left join (see Section 3.2.3), but it can also do inner joins by setting the argument `left = FALSE`. Like spatial subsetting,

---

[2]https://github.com/Robinlovelace/geocompr/blob/master/code/04-spatial-join.R

FIGURE 4.4: The spatial distribution of cycle hire points in London based on official data (blue) and OpenStreetMap data (red).

the default topological operator used by `st_join()` is `st_intersects()`. This can be changed with the `join` argument (see `?st_join` for details). In the example above, we have added features of a polygon layer to a point layer. In other cases, we might want to join point attributes to a polygon layer. There might be occasions where more than one point falls inside one polygon. In such a case `st_join()` duplicates the polygon feature: it creates a new row for each match.

4.2.4 Non-overlapping joins

Sometimes two geographic datasets do not touch but still have a strong geographic relationship enabling joins. The datasets `cycle_hire` and `cycle_hire_osm`, already attached in the **spData** package, provide a good example. Plotting them shows that they are often closely related but they do not touch, as shown in Figure 4.4, a base version of which is created with the following code below:

```
plot(st_geometry(cycle_hire), col = "blue")
plot(st_geometry(cycle_hire_osm), add = TRUE, pch = 3, col = "red")
```

We can check if any points are the same `st_intersects()` as shown below:

```
any(st_touches(cycle_hire, cycle_hire_osm, sparse = FALSE))
#> [1] FALSE
```

Imagine that we need to join the `capacity` variable in `cycle_hire_osm` onto the official 'target' data contained in `cycle_hire`. This is when a non-overlapping join is needed. The simplest method is to use the topological operator `st_is_within_distance()` shown in Section 4.2.2, using a threshold distance of 20 m. Note that, before performing the relation, both objects are transformed into a projected CRS. These projected objects are created below (note the affix `_P`, short for projected):

```
cycle_hire_P = st_transform(cycle_hire, 27700)
cycle_hire_osm_P = st_transform(cycle_hire_osm, 27700)
sel = st_is_within_distance(cycle_hire_P, cycle_hire_osm_P, dist = 20)
summary(lengths(sel) > 0)
#>      Mode    FALSE     TRUE
#> logical      304      438
```

This shows that there are 438 points in the target object `cycle_hire_P` within the threshold distance of `cycle_hire_osm_P`. How to retrieve the *values* associated with the respective `cycle_hire_osm_P` points? The solution is again with `st_join()`, but with an addition `dist` argument (set to 20 m below):

```
z = st_join(cycle_hire_P, cycle_hire_osm_P, st_is_within_distance, dist = 20)
nrow(cycle_hire)
#> [1] 742
nrow(z)
#> [1] 762
```

Note that the number of rows in the joined result is greater than the target. This is because some cycle hire stations in `cycle_hire_P` have multiple matches in `cycle_hire_osm_P`. To aggregate the values for the overlapping points and return the mean, we can use the aggregation methods learned in Chapter 3, resulting in an object with the same number of rows as the target:

```
z = z %>%
  group_by(id) %>%
  summarize(capacity = mean(capacity))
nrow(z) == nrow(cycle_hire)
#> [1] TRUE
```

The capacity of nearby stations can be verified by comparing a plot of the capacity of the source `cycle_hire_osm` data with the results in this new object (plots not shown):

```
plot(cycle_hire_osm["capacity"])
plot(z["capacity"])
```

The result of this join has used a spatial operation to change the attribute data associated with simple features; the geometry associated with each feature has remained unchanged.

4.2.5   Spatial data aggregation

Like attribute data aggregation, covered in Section 3.2.2, spatial data aggregation can be a way of *condensing* data. Aggregated data show some statistics about a variable (typically average or total) in relation to some kind of *grouping variable*. Section 3.2.2 demonstrated how `aggregate()` and `group_by() %>% summarize()` condense data based on attribute variables. This section demonstrates how the same functions work using spatial grouping variables.

Returning to the example of New Zealand, imagine you want to find out the average height of high points in each region. This is a good example of spatial aggregation: it is the geometry of the source (`y` or `nz` in this case) that defines how values in the target object (`x` or `nz_height`) are grouped. This is illustrated using the base `aggregate()` function below:

```
nz_avheight = aggregate(x = nz_height, by = nz, FUN = mean)
```

The result of the previous command is an `sf` object with the same geometry as the (spatial) aggregating object (`nz`).[3] The result of the previous operation is illustrated in Figure 4.5. The same result can also be generated using the 'tidy' functions `group_by()` and `summarize()` (used in combination with `st_join()`):

```
nz_avheight2 = nz %>%
  st_join(nz_height) %>%
  group_by(Name) %>%
  summarize(elevation = mean(elevation, na.rm = TRUE))
```

The resulting `nz_avheight` objects have the same geometry as the aggregating object `nz` but with a new column representing the mean average height of points within each region of New Zealand (other summary functions such as `median()` and `sd()` can be used in place of `mean()`). Note that regions containing no points have an associated `elevation` value of `NA`. For aggregating operations which also create new geometries, see Section 5.2.6.

Spatial congruence is an important concept related to spatial aggregation. An *aggregating object* (which we will refer to as `y`) is *congruent* with the target

---

[3]This can be verified with `identical(st_geometry(nz), st_geometry(nz_avheight))`.

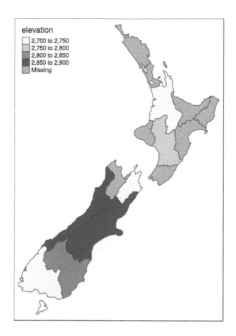

FIGURE 4.5: Average height of the top 101 high points across the regions of New Zealand.

object (x) if the two objects have shared borders. Often this is the case for administrative boundary data, whereby larger units — such as Middle Layer Super Output Areas (MSOAs[4]) in the UK or districts in many other European countries — are composed of many smaller units.

Incongruent aggregating objects, by contrast, do not share common borders with the target (Qiu et al., 2012). This is problematic for spatial aggregation (and other spatial operations) illustrated in Figure 4.6. Areal interpolation overcomes this issue by transferring values from one set of areal units to another. Algorithms developed for this task include area weighted and 'pycnophylactic' areal interpolation methods (Tobler, 1979).

The **spData** package contains a dataset named incongruent (colored polygons with black borders in the right panel of Figure 4.6) and a dataset named aggregating_zones (the two polygons with the translucent blue border in the right panel of Figure 4.6). Let us assume that the value column of incongruent refers to the total regional income in million Euros. How can we transfer the values of the underlying nine spatial polygons into the two polygons of aggregating_zones?

---

[4]https://www.ons.gov.uk/methodology/geography/ukgeographies/censusgeography

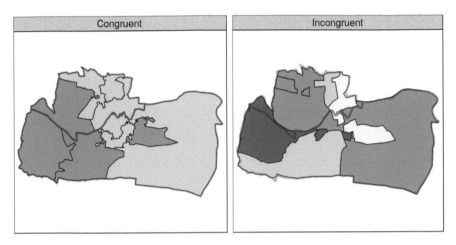

FIGURE 4.6: Illustration of congruent (left) and incongruent (right) areal units with respect to larger aggregating zones (translucent blue borders).

The simplest useful method for this is *area weighted* spatial interpolation. In this case values from the incongruent object are allocated to the aggregating_zones in proportion to area; the larger the spatial intersection between input and output features, the larger the corresponding value. For instance, if one intersection of incongruent and aggregating_zones is 1.5 km² but the whole incongruent polygon in question has 2 km² and a total income of 4 million Euros, then the target aggregating zone will obtain three quarters of the income, in this case 3 million Euros. This is implemented in st_interpolate_aw(), as demonstrated in the code chunk below.

```
agg_aw = st_interpolate_aw(incongruent[, "value"], aggregating_zones,
                           extensive = TRUE)
#> Warning in st_interpolate_aw(incongruent[, "value"], aggregating_zones, :
#> st_interpolate_aw assumes attributes are constant over areas of x
# show the aggregated result
agg_aw$value
#> [1] 19.6 25.7
```

In our case it is meaningful to sum up the values of the intersections falling into the aggregating zones since total income is a so-called spatially extensive variable. This would be different for spatially intensive variables, which are independent of the spatial units used, such as income per head or percentages[5]. In this case it is more meaningful to apply an average function when doing the aggregation instead of a sum function. To do so, one would only have to set the extensive parameter to FALSE.

---

[5]http://ibis.geog.ubc.ca/courses/geob370/notes/intensive_extensive.htm

4.2.6 Distance relations

While topological relations are binary — a feature either intersects with another or does not — distance relations are continuous. The distance between two objects is calculated with the st_distance() function. This is illustrated in the code chunk below, which finds the distance between the highest point in New Zealand and the geographic centroid of the Canterbury region, created in Section 4.2.1:

```
nz_heighest = nz_height %>% top_n(n = 1, wt = elevation)
canterbury_centroid = st_centroid(canterbury)
st_distance(nz_heighest, canterbury_centroid)
#> Units: [m]
#>         [,1]
#> [1,] 115540
```

There are two potentially surprising things about the result:

- It has units, telling us the distance is 100,000 meters, not 100,000 inches, or any other measure of distance.
- It is returned as a matrix, even though the result only contains a single value.

This second feature hints at another useful feature of st_distance(), its ability to return *distance matrices* between all combinations of features in objects x and y. This is illustrated in the command below, which finds the distances between the first three features in nz_height and the Otago and Canterbury regions of New Zealand represented by the object co.

```
co = filter(nz, grepl("Canter|Otag", Name))
st_distance(nz_height[1:3, ], co)
#> Units: [m]
#>         [,1]  [,2]
#> [1,] 123537 15498
#> [2,]  94283     0
#> [3,]  93019     0
```

Note that the distance between the second and third features in nz_height and the second feature in co is zero. This demonstrates the fact that distances between points and polygons refer to the distance to *any part of the polygon*: The second and third points in nz_height are *in* Otago, which can be verified by plotting them (result not shown):

```
plot(st_geometry(co)[2])
plot(st_geometry(nz_height)[2:3], add = TRUE)
```

4.3 Spatial operations on raster data

This section builds on Section 3.3, which highlights various basic methods for manipulating raster datasets, to demonstrate more advanced and explicitly spatial raster operations, and uses the objects elev and grain manually created in Section 3.3. For the reader's convenience, these datasets can be also found in the **spData** package.

4.3.1 Spatial subsetting

The previous chapter (Section 3.3) demonstrated how to retrieve values associated with specific cell IDs or row and column combinations. Raster objects can also be extracted by location (coordinates) and other spatial objects. To use coordinates for subsetting, one can 'translate' the coordinates into a cell ID with the **raster** function cellFromXY(). An alternative is to use raster::extract() (be careful, there is also a function called extract() in the **tidyverse**) to extract values. Both methods are demonstrated below to find the value of the cell that covers a point located 0.1 units from the origin.

```
id = cellFromXY(elev, xy = c(0.1, 0.1))
elev[id]
# the same as
raster::extract(elev, data.frame(x = 0.1, y = 0.1))
```

It is convenient that both functions also accept objects of class Spatial* Objects. Raster objects can also be subset with another raster object, as illustrated in Figure 4.7 (left panel) and demonstrated in the code chunk below:

```
clip = raster(xmn = 0.9, xmx = 1.8, ymn = -0.45, ymx = 0.45,
              res = 0.3, vals = rep(1, 9))
elev[clip]
#> [1] 18 24
# we can also use extract
# extract(elev, extent(clip))
```

Basically, this amounts to retrieving the values of the first raster (here: elev) falling within the extent of a second raster (here: clip).

So far, the subsetting returned the values of specific cells, however, when doing spatial subsetting, one often also expects a spatial object as an output. To do this, we can use again the [when we additionally set the drop parameter to FALSE. To illustrate this, we retrieve the first two cells of elev as an individual raster object. As mentioned in Section 3.3, the [operator accepts various

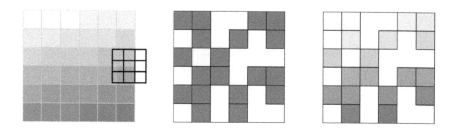

FIGURE 4.7: Subsetting raster values with the help of another raster (left). Raster mask (middle). Output of masking a raster (right).

inputs to subset rasters and returns a raster object when `drop = FALSE`. The code chunk below subsets the `elev` raster by cell ID and row-column index with identical results: the first two cells on the top row (only the first 2 lines of the output is shown):

```
elev[1:2, drop = FALSE]    # spatial subsetting with cell IDs
elev[1, 1:2, drop = FALSE] # spatial subsetting by row,column indices
#> class      : RasterLayer
#> dimensions : 1, 2, 2  (nrow, ncol, ncell)
#> ...
```

Another common use case of spatial subsetting is when a raster with `logical` (or `NA`) values is used to mask another raster with the same extent and resolution, as illustrated in Figure 4.7, middle and right panel. In this case, the `[`, `mask()` and `overlay()` functions can be used (results not shown):

```
# create raster mask
rmask = elev
values(rmask) = sample(c(NA, TRUE), 36, replace = TRUE)

# spatial subsetting
elev[rmask, drop = FALSE]          # with [ operator
mask(elev, rmask)                  # with mask()
overlay(elev, rmask, fun = "max")  # with overlay
```

In the code chunk above, we have created a mask object called `rmask` with values randomly assigned to `NA` and `TRUE`. Next, we want to keep those values of `elev` which are `TRUE` in `rmask`. In other words, we want to mask `elev` with `rmask`. These operations are in fact Boolean local operations since we compare cell-wise two rasters. The next subsection explores these and related operations in more detail.

4.3.2 Map algebra

Map algebra makes raster processing really fast. This is because raster datasets only implicitly store coordinates. To derive the coordinate of a specific cell, we have to calculate it using its matrix position and the raster resolution and origin. For the processing, however, the geographic position of a cell is barely relevant as long as we make sure that the cell position is still the same after the processing (one-to-one locational correspondence). Additionally, if two or more raster datasets share the same extent, projection and resolution, one could treat them as matrices for the processing. This is exactly what map algebra is doing in R. First, the **raster** package checks the headers of the rasters on which to perform any algebraic operation, and only if they are correspondent to each other, the processing goes on.[6] And secondly, map algebra retains the so-called one-to-one locational correspondence. This is where it substantially differs from matrix algebra which changes positions when for example multiplying or dividing matrices.

Map algebra (or cartographic modeling) divides raster operations into four subclasses (Tomlin, 1990), with each working on one or several grids simultaneously:

1. *Local* or per-cell operations.
2. *Focal* or neighborhood operations. Most often the output cell value is the result of a 3 x 3 input cell block.
3. *Zonal* operations are similar to focal operations, but the surrounding pixel grid on which new values are computed can have irregular sizes and shapes.
4. *Global* or per-raster operations; that means the output cell derives its value potentially from one or several entire rasters.

This typology classifies map algebra operations by the number/shape of cells used for each pixel processing step. For the sake of completeness, we should mention that raster operations can also be classified by discipline such as terrain, hydrological analysis or image classification. The following sections explain how each type of map algebra operations can be used, with reference to worked examples (also see `vignette("Raster")` for a technical description of map algebra).

---

[6]Map algebra operations are also possible with headerless rasters; in this case the user has to make sure that in fact there exists a one-to-one locational correspondence. An example showing how to import a headerless raster into R is provided in a post at `https://stat.ethz.ch/pipermail/r-sig-geo/2013-May/018278.html`.

4.3.3 Local operations

Local operations comprise all cell-by-cell operations in one or several layers. A good example is the classification of intervals of numeric values into groups such as grouping a digital elevation model into low (class 1), middle (class 2) and high elevations (class 3). Using the `reclassify()` command, we need first to construct a reclassification matrix, where the first column corresponds to the lower and the second column to the upper end of the class. The third column represents the new value for the specified ranges in column one and two. Here, we assign the raster values in the ranges 0–12, 12–24 and 24–36 are *reclassified* to take values 1, 2 and 3, respectively.

```
rcl = matrix(c(0, 12, 1, 12, 24, 2, 24, 36, 3), ncol = 3, byrow = TRUE)
recl = reclassify(elev, rcl = rcl)
```

We will perform several reclassifactions in Chapter 13.

Raster algebra is another classical use case of local operations. This includes adding, subtracting and squaring two rasters. Raster algebra also allows logical operations such as finding all raster cells that are greater than a specific value (5 in our example below). The **raster** package supports all these operations and more, as described in `vignette("Raster")` and demonstrated below (results not show):

```
elev + elev
elev^2
log(elev)
elev > 5
```

Instead of arithmetic operators, one can also use the `calc()` and `overlay()` functions. These functions are more efficient, hence, they are preferable in the presence of large raster datasets. Additionally, they allow you to directly store an output file.

The calculation of the normalized difference vegetation index (NDVI) is a well-known local (pixel-by-pixel) raster operation. It returns a raster with values between -1 and 1; positive values indicate the presence of living plants (mostly > 0.2). NDVI is calculated from red and near-infrared (NIR) bands of remotely sensed imagery, typically from satellite systems such as Landsat or Sentinel. Vegetation absorbs light heavily in the visible light spectrum, and especially in the red channel, while reflecting NIR light, explaining the NVDI formula:

$$NDVI = \frac{\text{NIR} - \text{Red}}{\text{NIR} + \text{Red}}$$

Predictive mapping is another interesting application of local raster operations.

The response variable corresponds to measured or observed points in space, for example, species richness, the presence of landslides, tree disease or crop yield. Consequently, we can easily retrieve space- or airborne predictor variables from various rasters (elevation, pH, precipitation, temperature, landcover, soil class, etc.). Subsequently, we model our response as a function of our predictors using `lm`, `glm`, `gam` or a machine-learning technique. Spatial predictions on raster objects can therefore be made by applying estimated coefficients to the predictor raster vaules, and summing the output raster values (see Chapter 14).

4.3.4 Focal operations

While local functions operate on one cell, though possibly from multiple layers, **focal** operations take into account a central cell and its neighbors. The neighborhood (also named kernel, filter or moving window) under consideration is typically of size 3-by-3 cells (that is the central cell and its eight surrounding neighbors), but can take on any other (not necessarily rectangular) shape as defined by the user. A focal operation applies an aggregation function to all cells within the specified neighborhood, uses the corresponding output as the new value for the the central cell, and moves on to the next central cell (Figure 4.8). Other names for this operation are spatial filtering and convolution (Burrough et al., 2015).

In R, we can use the `focal()` function to perform spatial filtering. We define the shape of the moving window with a `matrix` whose values correspond to weights (see `w` parameter in the code chunk below). Secondly, the `fun` parameter lets us specify the function we wish to apply to this neighborhood. Here, we choose the minimum, but any other summary function, including `sum()`, `mean()`, or `var()` can be used.

```
r_focal = focal(elev, w = matrix(1, nrow = 3, ncol = 3), fun = min)
```

We can quickly check if the output meets our expectations. In our example, the minimum value has to be always the upper left corner of the moving window (remember we have created the input raster by row-wise incrementing the cell values by one starting at the upper left corner). In this example, the weighting matrix consists only of 1s, meaning each cell has the same weight on the output, but this can be changed.

Focal functions or filters play a dominant role in image processing. Low-pass or smoothing filters use the mean function to remove extremes. In the case of categorical data, we can replace the mean with the mode, which is the most common value. By contrast, high-pass filters accentuate features. The line detection Laplace and Sobel filters might serve as an example here. Check the `focal()` help page for how to use them in R (this will also be used in the excercises at the end of this chapter).

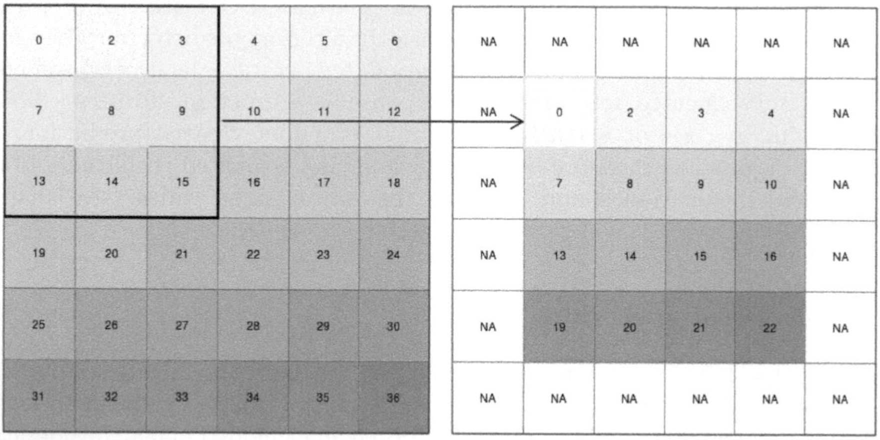

FIGURE 4.8: Input raster (left) and resulting output raster (right) due to a focal operation - finding the minimum value in 3-by-3 moving windows.

Terrain processing, the calculation of topographic characteristics such as slope, aspect and flow directions, relies on focal functions. `terrain()` can be used to calculate these metrics, although some terrain algorithms, including the Zevenbergen and Thorne method to compute slope, are not implemented in this **raster** function. Many other algorithms — including curvatures, contributing areas and wetness indices — are implemented in open source desktop geographic information system (GIS) software. Chapter 9 shows how to access such GIS functionality from within R.

4.3.5 Zonal operations

Zonal operations are similar to focal operations. The difference is that zonal filters can take on any shape instead of a predefined rectangular window. Our grain size raster is a good example (Figure 3.2) because the different grain sizes are spread in an irregular fashion throughout the raster.

To find the mean elevation for each grain size class, we can use the `zonal()` command. This kind of operation is also known as *zonal statistics* in the GIS world.

```
z = zonal(elev, grain, fun = "mean") %>%
  as.data.frame()
z
#>   zone mean
```

```
#> 1    1 17.8
#> 2    2 18.5
#> 3    3 19.2
```

This returns the statistics for each category, here the mean altitude for each grain size class, and can be added to the attribute table of the ratified raster (see previous chapter).

4.3.6 Global operations and distances

Global operations are a special case of zonal operations with the entire raster dataset representing a single zone. The most common global operations are descriptive statistics for the entire raster dataset such as the minimum or maximum (see Section 3.3.2). Aside from that, global operations are also useful for the computation of distance and weight rasters. In the first case, one can calculate the distance from each cell to a specific target cell. For example, one might want to compute the distance to the nearest coast (see also raster::distance()). We might also want to consider topography, that means, we are not only interested in the pure distance but would like also to avoid the crossing of mountain ranges when going to the coast. To do so, we can weight the distance with elevation so that each additional altitudinal meter 'prolongs' the Euclidean distance. Visibility and viewshed computations also belong to the family of global operations (in the exercises of Chapter 9, you will compute a viewshed raster).

Many map algebra operations have a counterpart in vector processing (Liu and Mason, 2009). Computing a distance raster (zonal operation) while only considering a maximum distance (logical focal operation) is the equivalent to a vector buffer operation (Section 5.2.5). Reclassifying raster data (either local or zonal function depending on the input) is equivalent to dissolving vector data (Section 4.2.3). Overlaying two rasters (local operation), where one contains NULL or NA values representing a mask, is similar to vector clipping (Section 5.2.5). Quite similar to spatial clipping is intersecting two layers (Section 4.2.1). The difference is that these two layers (vector or raster) simply share an overlapping area (see Figure 5.8 for an example). However, be careful with the wording. Sometimes the same words have slightly different meanings for raster and vector data models. Aggregating in the case of vector data refers to dissolving polygons, while it means increasing the resolution in the case of raster data. In fact, one could see dissolving or aggregating polygons as decreasing the resolution. However, zonal operations might be the better raster equivalent compared to changing the cell resolution. Zonal operations can dissolve the cells of one raster in accordance with the zones (categories) of another raster using an aggregation function (see above).

4.3.7 Merging rasters

Suppose we would like to compute the NDVI (see Section 4.3.3), and additionally want to compute terrain attributes from elevation data for observations within a study area. Such computations rely on remotely sensed information. The corresponding imagery is often divided into scenes covering a specific spatial extent. Frequently, a study area covers more than one scene. In these cases we would like to merge the scenes covered by our study area. In the easiest case, we can just merge these scenes, that is put them side by side. This is possible with digital elevation data (SRTM, ASTER). In the following code chunk we first download the SRTM elevation data for Austria and Switzerland (for the country codes, see the **raster** function ccodes()). In a second step, we merge the two rasters into one.

```
aut = getData("alt", country = "AUT", mask = TRUE)
ch = getData("alt", country = "CHE", mask = TRUE)
aut_ch = merge(aut, ch)
```

Raster's merge() command combines two images, and in case they overlap, it uses the value of the first raster. You can do exactly the same with gdalutils::mosaic_rasters() which is faster, and therefore recommended if you have to merge a multitude of large rasters stored on disk.

The merging approach is of little use when the overlapping values do not correspond to each other. This is frequently the case when you want to combine spectral imagery from scenes that were taken on different dates. The merge() command will still work but you will see a clear border in the resulting image. The mosaic() command lets you define a function for the overlapping area. For instance, we could compute the mean value. This might smooth the clear border in the merged result but it will most likely not make it disappear. To do so, we need a more advanced approach. Remote sensing scientists frequently apply histogram matching or use regression techniques to align the values of the first image with those of the second image. The packages **landsat** (histmatch(), relnorm(), PIF()), **satellite** (calcHistMatch()) and **RStoolbox** (histMatch(), pifMatch()) provide the corresponding functions. For a more detailed introduction on how to use R for remote sensing, we refer the reader to Wegmann et al. (2016).

4.4 Exercises

1. It was established in Section 4.2 that Canterbury was the region of New Zealand containing most of the 100 highest points in the

country. How many of these high points does the Canterbury region contain?

2. Which region has the second highest number of `nz_height` points in, and how many does it have?

3. Generalizing the question to all regions: how many of New Zealand's 16 regions contain points which belong to the top 100 highest points in the country? Which regions?

   - Bonus: create a table listing these regions in order of the number of points and their name.

4. Use `data(dem, package = "RQGIS")`, and reclassify the elevation in three classes: low, medium and high. Secondly, attach the NDVI raster (`data(ndvi, package = "RQGIS")`) and compute the mean NDVI and the mean elevation for each altitudinal class.

5. Apply a line detection filter to `raster(system.file("external/rlogo.grd", package = "raster"))`. Plot the result. Hint: Read `?raster::focal()`.

6. Calculate the NDVI of a Landsat image. Use the Landsat image provided by the **spDataLarge** package (`system.file("raster/landsat.tif", package="spDataLarge")`).

7. A StackOverflow post[7] shows how to compute distances to the nearest coastline using `raster::distance()`. Retrieve a digital elevation model of Spain, and compute a raster which represents distances to the coast across the country (hint: use `getData()`). Second, use a simple approach to weight the distance raster with elevation (other weighting approaches are possible, include flow direction and steepness); every 100 altitudinal meters should increase the distance to the coast by 10 km. Finally, compute the difference between the raster using the Euclidean distance and the raster weighted by elevation. Note: it may be wise to increase the cell size of the input raster to reduce compute time during this operation.

---

[7] https://stackoverflow.com/questions/35555709/global-raster-of-geographic-distances

5

Geometry operations

Prerequisites

- This chapter uses the same packages as Chapter 4 but with the addition of **spDataLarge**, which was installed in Chapter 2:

```
library(sf)
library(raster)
library(dplyr)
library(spData)
library(spDataLarge)
```

5.1 Introduction

The previous three chapters have demonstrated how geographic datasets are structured in R (Chapter 2) and how to manipulate them based on their non-geographic attributes (Chapter 3) and spatial properties (Chapter 4). This chapter extends these skills. After reading it — and attempting the exercises at the end — you should understand and have control over the geometry column in sf objects and the geographic location of pixels represented in rasters.

Section 5.2 covers transforming vector geometries with 'unary' and 'binary' operations. Unary operations work on a single geometry in isolation. This includes simplification (of lines and polygons), the creation of buffers and centroids, and shifting/scaling/rotating single geometries using 'affine transformations' (Sections 5.2.1 to 5.2.4). Binary transformations modify one geometry based on the shape of another. This includes clipping and geometry unions, covered in Sections 5.2.5 and 5.2.6, respectively. Type transformations (from a polygon to a line, for example) are demonstrated in Section 5.2.7.

Section 5.3 covers geometric transformations on raster objects. This involves changing the size and number of the underlying pixels, and assigning them new values. It teaches how to change the resolution (also called raster aggregation and disaggregation), the extent and the origin of a raster. These operations are especially useful if one would like to align raster datasets from diverse sources. Aligned raster objects share a one-to-one correspondence between pixels, allowing them to be processed using map algebra operations, described in Section 4.3.2. The final Section 5.4 connects vector and raster objects. It shows how raster values can be 'masked' and 'extracted' by vector geometries. Importantly it shows how to 'polygonize' rasters and 'rasterize' vector datasets, making the two data models more interchangeable.

5.2   Geometric operations on vector data

This section is about operations that in some way change the geometry of vector (sf) objects. It is more advanced than the spatial data operations presented in the previous chapter (in Section 4.2), because here we drill down into the geometry: the functions discussed in this section work on objects of class sfc in addition to objects of class sf.

5.2.1   Simplification

Simplification is a process for generalization of vector objects (lines and polygons) usually for use in smaller scale maps. Another reason for simplifying objects is to reduce the amount of memory, disk space and network bandwidth they consume: it may be wise to simplify complex geometries before publishing them as interactive maps. The **sf** package provides st_simplify(), which uses the GEOS implementation of the Douglas-Peucker algorithm to reduce the vertex count. st_simplify() uses the dTolerance to control the level of generalization in map units (see Douglas and Peucker, 1973, for details). Figure 5.1 illustrates simplification of a LINESTRING geometry representing the river Seine and tributaries. The simplified geometry was created by the following command:

```
seine_simp = st_simplify(seine, dTolerance = 2000)   # 2000 m
```

The resulting seine_simp object is a copy of the original seine but with fewer vertices. This is apparent, with the result being visually simpler (Figure 5.1, right) and consuming less memory than the original object, as verified below:

```
object.size(seine)
#> 17304 bytes
```

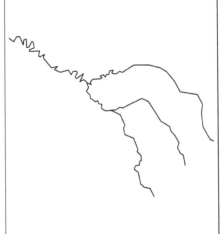

FIGURE 5.1: Comparison of the original and simplified geometry of the seine object.

```
object.size(seine_simp)
#> 8320 bytes
```

Simplification is also applicable for polygons. This is illustrated using `us_states`, representing the contiguous United States. As we show in Chapter 6, GEOS assumes that the data is in a projected CRS and this could lead to unexpected results when using a geographic CRS. Therefore, the first step is to project the data into some adequate projected CRS, such as US National Atlas Equal Area (epsg = 2163) (on the left in Figure 5.2):

```
us_states2163 = st_transform(us_states, 2163)
```

`st_simplify()` works equally well with projected polygons:

```
us_states_simp1 = st_simplify(us_states2163, dTolerance = 100000)   # 100 km
```

A limitation with `st_simplify()` is that it simplifies objects on a per-geometry basis. This means the 'topology' is lost, resulting in overlapping and 'holy' areal units illustrated in Figure 5.2 (middle panel). `ms_simplify()` from **rmapshaper** provides an alternative that overcomes this issue. By default it uses the Visvalingam algorithm, which overcomes some limitations of the Douglas-Peucker algorithm (Visvalingam and Whyatt, 1993). The following code chunk uses

FIGURE 5.2: Polygon simplification in action, comparing the original geometry of the contiguous United States with simplified versions, generated with functions from sf (center) and rmapshaper (right) packages.

this function to simplify `us_states2163`. The result has only 1% of the vertices of the input (set using the argument `keep`) but its number of objects remains intact because we set `keep_shapes = TRUE`:[1]

```
# proportion of points to retain (0-1; default 0.05)
us_states2163$AREA = as.numeric(us_states2163$AREA)
us_states_simp2 = rmapshaper::ms_simplify(us_states2163, keep = 0.01,
                                           keep_shapes = TRUE)
```

Finally, the visual comparison of the original dataset and the two simplified versions shows differences between the Douglas-Peucker (`st_simplify`) and Visvalingam (`ms_simplify`) algorithm outputs (Figure 5.2):

5.2.2 Centroids

Centroid operations identify the center of geographic objects. Like statistical measures of central tendency (including mean and median definitions of 'average'), there are many ways to define the geographic center of an object. All of them create single point representations of more complex vector objects.

The most commonly used centroid operation is the *geographic centroid*. This type of centroid operation (often referred to as 'the centroid') represents the center of mass in a spatial object (think of balancing a plate on your finger). Geographic centroids have many uses, for example to create a simple point representation of complex geometries, or to estimate distances between polygons. They can be calculated with the **sf** function `st_centroid()` as demonstrated in the code below, which generates the geographic centroids of regions in New Zealand and tributaries to the River Seine, illustrated with black points in Figure 5.3.

---

[1]Simplification of multipolygon objects can remove small internal polygons, even if the `keep_shapes` argument is set to TRUE. To prevent this, you need to set `explode = TRUE`. This option converts all mutlipolygons into separate polygons before its simplification.

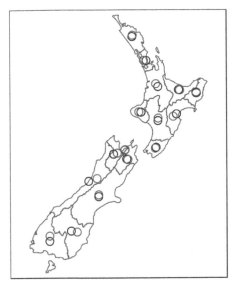
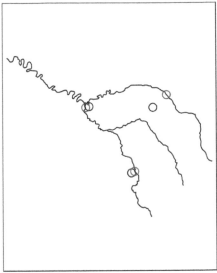

FIGURE 5.3: Centroids (black points) and 'points on surface' (red points) of New Zealand's regions (left) and the Seine (right) datasets.

```
nz_centroid = st_centroid(nz)
seine_centroid = st_centroid(seine)
```

Sometimes the geographic centroid falls outside the boundaries of their parent objects (think of a doughnut). In such cases *point on surface* operations can be used to guarantee the point will be in the parent object (e.g., for labeling irregular multipolygon objects such as island states), as illustrated by the red points in Figure 5.3. Notice that these red points always lie on their parent objects. They were created with st_point_on_surface() as follows:[2]

```
nz_pos = st_point_on_surface(nz)
seine_pos = st_point_on_surface(seine)
```

Other types of centroids exist, including the *Chebyshev center* and the *visual center*. We will not explore these here but it is possible to calculate them using R, as we'll see in Chapter 10.

---

[2]A description of how st_point_on_surface() works is provided at https://gis.stackexchange.com/q/76498.

5 km buffer   50 km buffer

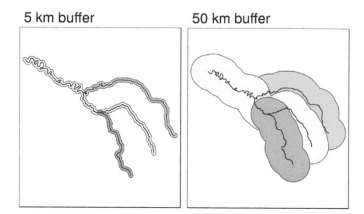

FIGURE 5.4: Buffers around the Seine dataset of 5 km (left) and 50 km (right). Note the colors, which reflect the fact that one buffer is created per geometry feature.

5.2.3   Buffers

Buffers are polygons representing the area within a given distance of a geometric feature: regardless of whether the input is a point, line or polygon, the output is a polygon. Unlike simplification (which is often used for visualization and reducing file size) buffering tends to be used for geographic data analysis. How many points are within a given distance of this line? Which demographic groups are within travel distance of this new shop? These kinds of questions can be answered and visualized by creating buffers around the geographic entities of interest.

Figure 5.4 illustrates buffers of different sizes (5 and 50 km) surrounding the river Seine and tributaries. These buffers were created with commands below, which show that the command `st_buffer()` requires at least two arguments: an input geometry and a distance, provided in the units of the CRS (in this case meters):

```
seine_buff_5km = st_buffer(seine, dist = 5000)
seine_buff_50km = st_buffer(seine, dist = 50000)
```

The third and final argument of `st_buffer()` is `nQuadSegs`, which means 'number of segments per quadrant' and is set by default to 30 (meaning circles created by buffers are composed of $4 \times 30 = 120$ lines). This argument rarely needs to be set. Unusual cases where it may be useful include when the memory consumed by the output of a buffer operation is a major concern (in which case it should be reduced) or when very high precision is needed (in which case it should be increased).

5.2.4 Affine transformations

Affine transformation is any transformation that preserves lines and parallelism. However, angles or length are not necessarily preserved. Affine transformations include, among others, shifting (translation), scaling and rotation. Additionally, it is possible to use any combination of these. Affine transformations are an essential part of geocomputation. For example, shifting is needed for labels placement, scaling is used in non-contiguous area cartograms (see Section 8.6), and many affine transformations are applied when reprojecting or improving the geometry that was created based on a distorted or wrongly projected map. The **sf** package implements affine transformation for objects of classes sfg and sfc.

```
nz_sfc = st_geometry(nz)
```

Shifting moves every point by the same distance in map units. It could be done by adding a numerical vector to a vector object. For example, the code below shifts all y-coordinates by 100,000 meters to the north, but leaves the x-coordinates untouched (left panel of Figure 5.5).

```
nz_shift = nz_sfc + c(0, 100000)
```

Scaling enlarges or shrinks objects by a factor. It can be applied either globally or locally. Global scaling increases or decreases all coordinates values in relation to the origin coordinates, while keeping all geometries topological relations intact. It can be done by subtraction or multiplication of asfg or sfc object.

Local scaling treats geometries independently and requires points around which geometries are going to be scaled, e.g., centroids. In the example below, each geometry is shrunk by a factor of two around the centroids (middle panel in Figure 5.5). To achieve that, each object is firstly shifted in a way that its center has coordinates of 0, 0 ((nz_sfc - nz_centroid_sfc)). Next, the sizes of the geometries are reduced by half (* 0.5). Finally, each object's centroid is moved back to the input data coordinates (+ nz_centroid_sfc).

```
nz_centroid_sfc = st_centroid(nz_sfc)
nz_scale = (nz_sfc - nz_centroid_sfc) * 0.5 + nz_centroid_sfc
```

Rotation of two-dimensional coordinates requires a rotation matrix:

Shift   Scale   Rotate

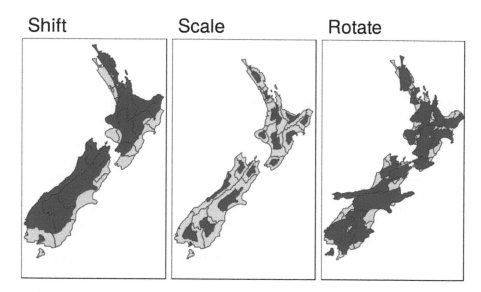

FIGURE 5.5: Illustrations of affine transformations: shift, scale and rotate.

$$R = \begin{bmatrix} \cos\theta & -\sin\theta \\ \sin\theta & \cos\theta \end{bmatrix}$$

It rotates points in a counterclockwise direction. The rotation matrix can be implemented in R as:

```
rotation = function(a){
  r = a * pi / 180 #degrees to radians
  matrix(c(cos(r), sin(r), -sin(r), cos(r)), nrow = 2, ncol = 2)
}
```

The rotation function accepts one argument a - a rotation angle in degrees. Rotation could be done around selected points, such as centroids (right panel of Figure 5.5). See vignette("sf3") for more examples.

```
nz_rotate = (nz_sfc - nz_centroid_sfc) * rotation(30) + nz_centroid_sfc
```

Finally, the newly created geometries can replace the old ones with the st_set_geometry() function:

```
nz_scale_sf = st_set_geometry(nz, nz_scale)
```

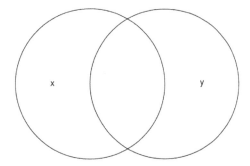

FIGURE 5.6: Overlapping circles.

5.2.5 Clipping

Spatial clipping is a form of spatial subsetting that involves changes to the `geometry` columns of at least some of the affected features.

Clipping can only apply to features more complex than points: lines, polygons and their 'multi' equivalents. To illustrate the concept we will start with a simple example: two overlapping circles with a center point one unit away from each other and a radius of one (Figure 5.6).

```
b = st_sfc(st_point(c(0, 1)), st_point(c(1, 1))) # create 2 points
b = st_buffer(b, dist = 1) # convert points to circles
plot(b)
text(x = c(-0.5, 1.5), y = 1, labels = c("x", "y")) # add text
```

Imagine you want to select not one circle or the other, but the space covered by both x *and* y. This can be done using the function `st_intersection()`, illustrated using objects named x and y which represent the left- and right-hand circles (Figure 5.7).

```
x = b[1]
y = b[2]
x_and_y = st_intersection(x, y)
plot(b)
plot(x_and_y, col = "lightgrey", add = TRUE) # color intersecting area
```

The subsequent code chunk demonstrates how this works for all combinations of the 'Venn' diagram representing x and y, inspired by Figure 5.1[3] of the book *R for Data Science* (Grolemund and Wickham, 2016).

To illustrate the relationship between subsetting and clipping spatial data,

---

[3]http://r4ds.had.co.nz/transform.html#logical-operators

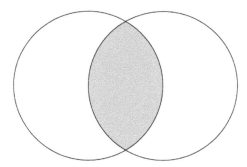

FIGURE 5.7: Overlapping circles with a gray color indicating intersection between them.

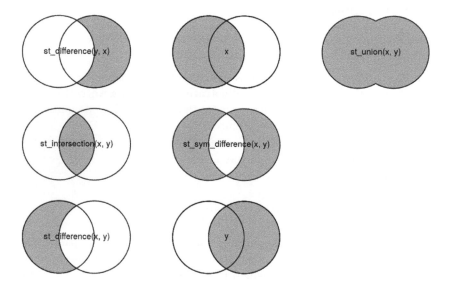

FIGURE 5.8: Spatial equivalents of logical operators.

we will subset points that cover the bounding box of the circles x and y in Figure 5.8. Some points will be inside just one circle, some will be inside both and some will be inside neither. st_sample() is used below to generate a *simple random* distribution of points within the extent of circles x and y, resulting in output illustrated in Figure 5.9.

```
bb = st_bbox(st_union(x, y))
box = st_as_sfc(bb)
set.seed(2017)
p = st_sample(x = box, size = 10)
```

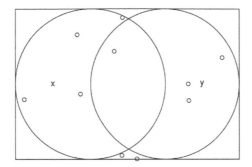

FIGURE 5.9: Randomly distributed points within the bounding box enclosing circles x and y.

```
plot(box)
plot(x, add = TRUE)
plot(y, add = TRUE)
plot(p, add = TRUE)
text(x = c(-0.5, 1.5), y = 1, labels = c("x", "y"))
```

The logical operator way would find the points inside both x and y using a spatial predicate such as st_intersects(), whereas the intersection method simply finds the points inside the intersecting region created above as x_and_y. As demonstrated below the results are identical, but the method that uses the clipped polygon is more concise:

```
sel_p_xy = st_intersects(p, x, sparse = FALSE)[, 1] &
  st_intersects(p, y, sparse = FALSE)[, 1]
p_xy1 = p[sel_p_xy]
p_xy2 = p[x_and_y]
identical(p_xy1, p_xy2)
#> [1] TRUE
```

5.2.6 Geometry unions

As we saw in Section 3.2.2, spatial aggregation can silently dissolve the geometries of touching polygons in the same group. This is demonstrated in the code chunk below in which 49 us_states are aggregated into 4 regions using base and **tidyverse** functions (see results in Figure 5.10):

```
regions = aggregate(x = us_states[, "total_pop_15"], by = list(us_states$REGION),
                    FUN = sum, na.rm = TRUE)
```

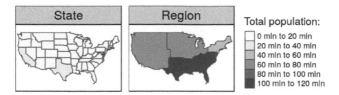

FIGURE 5.10: Spatial aggregation on contiguous polygons, illustrated by aggregating the population of US states into regions, with population represented by color. Note the operation automatically dissolves boundaries between states.

```
regions2 = us_states %>% group_by(REGION) %>%
  summarize(pop = sum(total_pop_15, na.rm = TRUE))
```

What is going on in terms of the geometries? Behind the scenes, both `aggregate()` and `summarize()` combine the geometries and dissolve the boundaries between them using `st_union()`. This is demonstrated in the code chunk below which creates a united western US:

```
us_west = us_states[us_states$REGION == "West", ]
us_west_union = st_union(us_west)
```

The function can take two geometries and unite them, as demonstrated in the code chunk below which creates a united western block incorporating Texas (challenge: reproduce and plot the result):

```
texas = us_states[us_states$NAME == "Texas", ]
texas_union = st_union(us_west_union, texas)
```

5.2.7   Type transformations

Geometry casting is a powerful operation that enables transformation of the geometry type. It is implemented in the `st_cast` function from the **sf** package. Importantly, `st_cast` behaves differently on single simple feature geometry (`sfg`) objects, simple feature geometry column (`sfc`) and simple features objects.

Let's create a multipoint to illustrate how geometry casting works on simple feature geometry (`sfg`) objects:

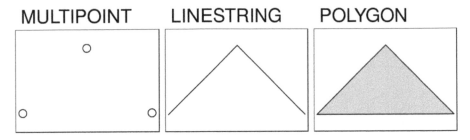

FIGURE 5.11: Examples of linestring and polygon casted from a multipoint geometry.

```
multipoint = st_multipoint(matrix(c(1, 3, 5, 1, 3, 1), ncol = 2))
```

In this case, `st_cast` can be useful to transform the new object into linestring or polygon (Figure 5.11):

```
linestring = st_cast(multipoint, "LINESTRING")
polyg = st_cast(multipoint, "POLYGON")
```

Conversion from multipoint to linestring is a common operation that creates a line object from ordered point observations, such as GPS measurements or geotagged media. This allows spatial operations such as the length of the path traveled. Conversion from multipoint or linestring to polygon is often used to calculate an area, for example from the set of GPS measurements taken around a lake or from the corners of a building lot.

The transformation process can be also reversed using `st_cast`:

```
multipoint_2 = st_cast(linestring, "MULTIPOINT")
multipoint_3 = st_cast(polyg, "MULTIPOINT")
all.equal(multipoint, multipoint_2, multipoint_3)
#> [1] TRUE
```

For single simple feature geometries (`sfg`), `st_cast` also provides geometry casting from non-multi-types to multi-types (e.g., POINT to MULTIPOINT) and from multi-types to non-multi-types. However, only the first element of the old object would remain in the second group of cases.

Geometry casting of simple features geometry column (`sfc`) and simple features objects works the same as for single geometries in most of the cases. One important difference is the conversion between multi-types to non-multi-types. As a result of this process, multi-objects are split into many non-multi-objects.

TABLE 5.1: Geometry casting on simple feature geometries (see Section 2.1) with input type by row and output type by column

	POI	MPOI	LIN	MLIN	POL	MPOL	GC
POI(1)	1	1	1	NA	NA	NA	NA
MPOI(1)	4	1	1	1	1	NA	NA
LIN(1)	5	1	1	1	1	NA	NA
MLIN(1)	7	2	2	1	NA	NA	NA
POL(1)	5	1	1	1	1	1	NA
MPOL(1)	10	1	NA	1	2	1	1
GC(1)	9	1	NA	NA	NA	NA	1

Note: Values like (1) represent the number of features; NA means the operation is not possible. Abbreviations: POI, LIN, POL and GC refer to POINT, LINESTRING, POLYGON and GEOMETRYCOLLECTION. The MULTI version of these geometry types is indicated by a preceding M, e.g., MPOI is the acronym for MULTIPOINT.

Table 5.1 shows possible geometry type transformations on simple feature objects. Each input simple feature object with only one element (first column) is transformed directly into another geometry type. Several of the transformations are not possible, for example, you cannot convert a single point into a multilinestring or a polygon (so the cells [1, 4:5] in the table are NA). On the other hand, some of the transformations are splitting the single element input object into a multi-element object. You can see that, for example, when you cast a multipoint consisting of five pairs of coordinates into a point.

Let's try to apply geometry type transformations on a new object, `multilinestring_sf`, as an example (on the left in Figure 5.12):

```
multilinestring_list = list(matrix(c(1, 4, 5, 3), ncol = 2),
                            matrix(c(4, 4, 4, 1), ncol = 2),
                            matrix(c(2, 4, 2, 2), ncol = 2))
multilinestring = st_multilinestring((multilinestring_list))
multilinestring_sf = st_sf(geom = st_sfc(multilinestring))
multilinestring_sf
#> Simple feature collection with 1 feature and 0 fields
#> geometry type:   MULTILINESTRING
#> dimension:       XY
#> bbox:            xmin: 1 ymin: 1 xmax: 4 ymax: 5
#> epsg (SRID):     NA
#> proj4string:     NA
```

MULTILINESTRING   LINESTRING

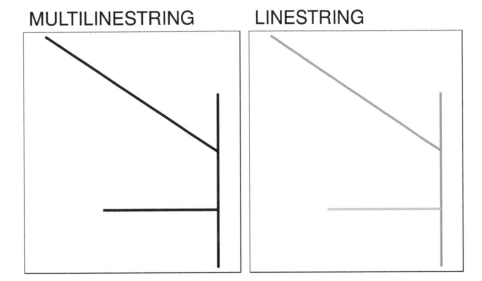

FIGURE 5.12: Examples of type casting between MULTILINESTRING (left) and LINESTRING (right).

```
#>                              geom
#> 1 MULTILINESTRING ((1 5, 4 3)...
```

You can imagine it as a road or river network. The new object has only one row that defines all the lines. This restricts the number of operations that can be done, for example it prevents adding names to each line segment or calculating lengths of single lines. The st_cast function can be used in this situation, as it separates one mutlilinestring into three linestrings:

```
linestring_sf2 = st_cast(multilinestring_sf, "LINESTRING")
linestring_sf2
#> Simple feature collection with 3 features and 0 fields
#> geometry type:  LINESTRING
#> dimension:      XY
#> bbox:           xmin: 1 ymin: 1 xmax: 4 ymax: 5
#> epsg (SRID):    NA
#> proj4string:    NA
#>                 geom
#> 1 LINESTRING (1 5, 4 3)
#> 2 LINESTRING (4 4, 4 1)
#> 3 LINESTRING (2 2, 4 2)
```

The newly created object allows for attributes creation (see more in Section 3.2.4) and length measurements:

```
linestring_sf2$name = c("Riddle Rd", "Marshall Ave", "Foulke St")
linestring_sf2$length = st_length(linestring_sf2)
linestring_sf2
#> Simple feature collection with 3 features and 2 fields
#> geometry type:  LINESTRING
#> dimension:      XY
#> bbox:           xmin: 1 ymin: 1 xmax: 4 ymax: 5
#> epsg (SRID):    NA
#> proj4string:    NA
#>                   geom            name length
#> 1 LINESTRING (1 5, 4 3)      Riddle Rd   3.61
#> 2 LINESTRING (4 4, 4 1) Marshall Ave    3.00
#> 3 LINESTRING (2 2, 4 2)      Foulke St   2.00
```

5.3   Geometric operations on raster data

Geometric raster operations include the shift, flipping, mirroring, scaling, rotation or warping of images. These operations are necessary for a variety of applications including georeferencing, used to allow images to be overlaid on an accurate map with a known CRS (Liu and Mason, 2009). A variety of georeferencing techniques exist, including:

*   Georeferencing based on known ground control points[4].
*   Orthorectification also georeferences an image, but additionally takes into account local topography.
*   Image (co-)registration is the process of aligning one image with another (in terms of coordinate reference system, origin and resolution). Registration becomes necessary for images from the same scene but shot from different sensors or from different angles or at different points in time.

R is unsuitable for the first two points since these often require manual intervention which is why they are usually done with the help of dedicated GIS software (see also Chapter 9). On the other hand, aligning several images is possible in R and this section shows among others how to do so. This often includes changing the extent, the resolution and the origin of an image. A matching projection is of course also required but is already covered in Section 6.6. In any case, there are other reasons to perform a geometric operation

---

[4]http://www.qgistutorials.com/en/docs/georeferencing_basics.html

on a single raster image. For instance, in Chapter 13 we define metropolitan areas in Germany as 20 km^2 pixels with more than 500,000 inhabitants. The original inhabitant raster, however, has a resolution of 1 km^2 which is why we will decrease (aggregate) the resolution by a factor of 20 (see Section 13.5). Another reason for aggregating a raster is simply to decrease run-time or save disk space. Of course, this is only possible if the task at hand allows a coarser resolution. Sometimes a coarser resolution is sufficient for the task at hand.

5.3.1 Geometric intersections

In Section 4.3.1 we have shown how to extract values from a raster overlaid by other spatial objects. To retrieve a spatial output, we can use almost the same subsetting syntax. The only difference is that we have to make clear that we would like to keep the matrix structure by setting the drop-parameter to FALSE. This will return a raster object containing the cells whose midpoints overlap with clip.

```
data("elev", package = "spData")
clip = raster(xmn = 0.9, xmx = 1.8, ymn = -0.45, ymx = 0.45,
              res = 0.3, vals = rep(1, 9))
elev[clip, drop = FALSE]
#> class        : RasterLayer
#> dimensions   : 2, 1, 2  (nrow, ncol, ncell)
#> resolution   : 0.5, 0.5  (x, y)
#> extent       : 1, 1.5, -0.5, 0.5  (xmin, xmax, ymin, ymax)
#> coord. ref.  : +proj=longlat +datum=WGS84 +ellps=WGS84 +towgs84=0,0,0
#> data source  : in memory
#> names        : layer
#> values       : 18, 24  (min, max)
```

For the same operation we can also use the intersect() and crop() command.

5.3.2 Extent and origin

When merging or performing map algebra on rasters, their resolution, projection, origin and/or extent have to match. Otherwise, how should we add the values of one raster with a resolution of 0.2 decimal degrees to a second with a resolution of 1 decimal degree? The same problem arises when we would like to merge satellite imagery from different sensors with different projections and resolutions. We can deal with such mismatches by aligning the rasters.

In the simplest case, two images only differ with regard to their extent. Following code adds one row and two columns to each side of the raster while setting all new values to an elevation of 1000 meters (Figure 5.13).

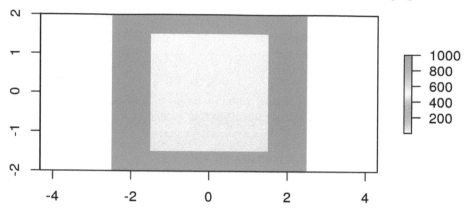

FIGURE 5.13: Original raster extended by one row on each side (top, bottom) and two columns on each side (right, left).

```
data(elev, package = "spData")
elev_2 = extend(elev, c(1, 2), value = 1000)
plot(elev_2)
```

Performing an algebraic operation on two objects with differing extents in R, the **raster** package returns the result for the intersection, and says so in a warning.

```
elev_3 = elev + elev_2
#> Warning in elev + elev_2: Raster objects have different extents. Result for
#> their intersection is returned
```

However, we can also align the extent of two rasters with extend(). Instead of telling the function how many rows or columns should be added (as done before), we allow it to figure it out by using another raster object. Here, we extend the elev object to the extent of elev_2. The newly added rows and column receive the default value of the value parameter, i.e., NA.

```
elev_4 = extend(elev, elev_2)
```

The origin is the point closest to (0, 0) if you moved towards it (starting from any given cell corner of the raster) in steps of x and y resolution.

```
origin(elev_4)
#> [1] 0 0
```

If two rasters have different origins, their cells do not overlap completely which

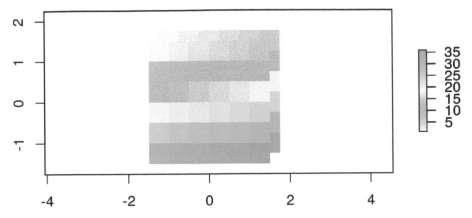

FIGURE 5.14: Rasters with identical values but different origins.

would make map algebra impossible. To change the origin , use `origin()`.[5]
Looking at Figure 5.14 reveals the effect of changing the origin.

```
# change the origin
origin(elev_4) = c(0.25, 0.25)
plot(elev_4)
# and add the original raster
plot(elev, add = TRUE)
```

Note that changing the resolution frequently (next section) also changes the origin.

5.3.3 Aggregation and disaggregation

Raster datasets can also differ with regard to their resolution. To match resolutions, one can either decrease (`aggregate()`) or increase (`disaggregate()`) the resolution of one raster.[6] As an example, we here change the spatial resolution of `dem` (found in the **RQGIS** package) by a factor of 5 (Figure 5.15). Additionally, the output cell value should correspond to the mean of the input cells (note that one could use other functions as well, such as `median()`, `sum()`, etc.):

---

[5]If the origins of two raster datasets are just marginally apart, it sometimes is sufficient to simply increase the `tolerance` argument of `raster::rasterOptions()`.

[6]Here we refer to spatial resolution. In remote sensing the spectral (spectral bands), temporal (observations through time of the same area) and radiometric (color depth) resolution are also important. Check out the `stackApply()` example in the documentation for getting an idea on how to do temporal raster aggregation.

Original

Aggregated

FIGURE 5.15: Original raster (left). Aggregated raster (right).

```
data("dem", package = "RQGIS")
dem_agg = aggregate(dem, fact = 5, fun = mean)
```

By contrast, the `disaggregate()` function increases the resolution. However, we have to specify a method on how to fill the new cells. The `disaggregate()` function provides two methods. The first (nearest neighbor, `method = ""`) simply gives all output cells the value of the nearest input cell, and hence duplicates values which leads to a blocky output image.

The `bilinear` method, in turn, is an interpolation technique that uses the four nearest pixel centers of the input image (salmon colored points in Figure 5.16) to compute an average weighted by distance (arrows in Figure 5.16 as the value of the output cell - square in the upper left corner in Figure 5.16).

```
dem_disagg = disaggregate(dem_agg, fact = 5, method = "bilinear")
identical(dem, dem_disagg)
#> [1] FALSE
```

Comparing the values of `dem` and `dem_disagg` tells us that they are not identical (you can also use `compareRaster()` or `all.equal()`). However, this was hardly to be expected, since disaggregating is a simple interpolation technique. It is important to keep in mind that disaggregating results in a finer resolution; the corresponding values, however, are only as accurate as their lower resolution source.

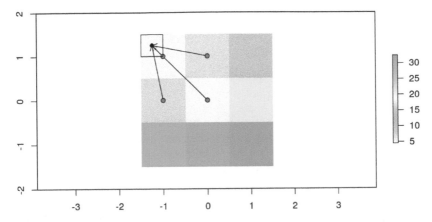

FIGURE 5.16: Bilinear disaggregation in action.

The process of computing values for new pixel locations is also called resampling. In fact, the **raster** package provides a `resample()` function. It lets you align several raster properties in one go, namely origin, extent and resolution. By default, it uses the `bilinear`-interpolation.

```
# add 2 rows and columns, i.e. change the extent
dem_agg = extend(dem_agg, 2)
dem_disagg_2 = resample(dem_agg, dem)
```

Finally, in order to align many (possibly hundreds or thousands of) images stored on disk, you could use the `gdalUtils::align_rasters()` function. However, you may also use **raster** with very large datasets. This is because **raster**:

1. Lets you work with raster datasets that are too large to fit into the main memory (RAM) by only processing chunks of it.
2. Tries to facilitate parallel processing. For more information, see the help pages of `beginCluster()` and `clusteR()`. Additionally, check out the *Multi-core functions* section in `vignette("functions", package = "raster")`.

5.4 Raster-vector interactions

This section focuses on interactions between raster and vector geographic data models, introduced in Chapter 2. It includes four main techniques: raster cropping and masking using vector objects (Section 5.4.1); extracting raster values using different types of vector data (Section 5.4.2); and raster-vector conversion (Sections 5.4.3 and 5.4.4). The above concepts are demonstrated using data used in previous chapters to understand their potential real-world applications.

5.4.1 Raster cropping

Many geographic data projects involve integrating data from many different sources, such as remote sensing images (rasters) and administrative boundaries (vectors). Often the extent of input raster datasets is larger than the area of interest. In this case raster **cropping** and **masking** are useful for unifying the spatial extent of input data. Both operations reduce object memory use and associated computational resources for subsequent analysis steps, and may be a necessary preprocessing step before creating attractive maps involving raster data.

We will use two objects to illustrate raster cropping:

- A raster object srtm representing elevation (meters above sea level) in southwestern Utah.
- A vector (sf) object zion representing Zion National Park.

Both target and cropping objects must have the same projection. The following code chunk therefore not only loads the datasets, from the **spDataLarge** package installed in Chapter 2, it also reprojects zion (see Section 6 for more on reprojection):

```
srtm = raster(system.file("raster/srtm.tif", package = "spDataLarge"))
zion = st_read(system.file("vector/zion.gpkg", package = "spDataLarge"))
zion = st_transform(zion, projection(srtm))
```

We will use crop() from the **raster** package to crop the srtm raster. crop() reduces the rectangular extent of the object passed to its first argument based on the extent of the object passed to its second argument, as demonstrated in the command below (which generates Figure 5.17(B) — note the smaller extent of the raster background):

```
srtm_cropped = crop(srtm, zion)
```

Related to crop() is the **raster** function mask(), which sets values outside of the bounds of the object passed to its second argument to NA. The following command therefore masks every cell outside of the Zion National Park boundaries (Figure 5.17(C)):

```
srtm_masked = mask(srtm, zion)
```

Changing the settings of mask() yields different results. Setting maskvalue = 0,

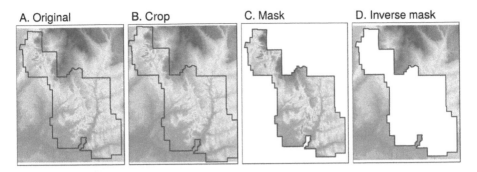

FIGURE 5.17: Illustration of raster cropping and raster masking.

for example, will set all pixels outside the national park to 0. Setting inverse = TRUE will mask everything *inside* the bounds of the park (see ?mask for details) (Figure 5.17(D)).

```
srtm_inv_masked = mask(srtm, zion, inverse = TRUE)
```

5.4.2 Raster extraction

Raster extraction is the process of identifying and returning the values associated with a 'target' raster at specific locations, based on a (typically vector) geographic 'selector' object. The results depend on the type of selector used (points, lines or polygons) and arguments passed to the raster::extract() function, which we use to demonstrate raster extraction. The reverse of raster extraction — assigning raster cell values based on vector objects — is rasterization, described in Section 5.4.3.

The simplest example is extracting the value of a raster cell at specific **points**. For this purpose, we will use zion_points, which contain a sample of 30 locations within the Zion National Park (Figure 5.18). The following command extracts elevation values from srtm and assigns the resulting vector to a new column (elevation) in the zion_points dataset:

```
data("zion_points", package = "spDataLarge")
zion_points$elevation = raster::extract(srtm, zion_points)
```

The buffer argument can be used to specify a buffer radius (in meters) around each point. The result of raster::extract(srtm, zion_points, buffer = 1000), for example, is a list of vectors, each of which representing the values of cells inside the buffer associated with each point. In practice, this example is a special case of extraction with a polygon selector, described below.

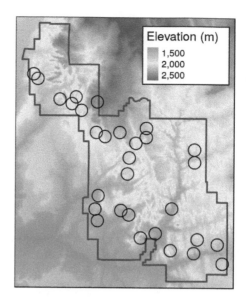

FIGURE 5.18: Locations of points used for raster extraction.

Raster extraction also works with **line** selectors. To demonstrate this, the code below creates zion_transect, a straight line going from northwest to southeast of the Zion National Park, illustrated in Figure 5.19(A) (see Section 2.2 for a recap on the vector data model):

```
zion_transect = cbind(c(-113.2, -112.9), c(37.45, 37.2)) %>%
  st_linestring() %>%
  st_sfc(crs = projection(srtm)) %>%
  st_sf()
```

The utility of extracting heights from a linear selector is illustrated by imagining that you are planning a hike. The method demonstrated below provides an 'elevation profile' of the route (the line does not need to be straight), useful for estimating how long it will take due to long climbs:

```
transect = raster::extract(srtm, zion_transect,
                           along = TRUE, cellnumbers = TRUE)
```

Note the use of along = TRUE and cellnumbers = TRUE arguments to return cell IDs *along* the path. The result is a list containing a matrix of cell IDs in the first column and elevation values in the second. The number of list elements is equal to the number of lines or polygons from which we are extracting values. The subsequent code chunk first converts this tricky matrix-in-a-list object into a

A. Line extraction

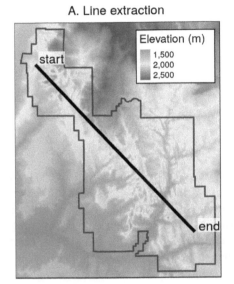

B. Elevation along the line

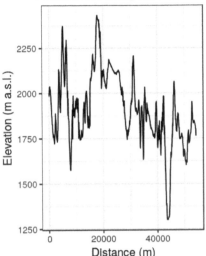

FIGURE 5.19: Location of a line used for raster extraction (left) and the elevation along this line (right).

simple data frame, returns the coordinates associated with each extracted cell, and finds the associated distances along the transect (see `?geosphere::distGeo()` for details):

```
transect_df = purrr::map_dfr(transect, as_data_frame, .id = "ID")
transect_coords = xyFromCell(srtm, transect_df$cell)
transect_df$dist = c(0, cumsum(geosphere::distGeo(transect_coords)))
```

The resulting `transect_df` can be used to create elevation profiles, as illustrated in Figure 5.19(B).

The final type of geographic vector object for raster extraction is **polygons**. Like lines and buffers, polygons tend to return many raster values per polygon. This is demonstrated in the command below, which results in a data frame with column names `ID` (the row number of the polygon) and `srtm` (associated elevation values):

```
zion_srtm_values = raster::extract(x = srtm, y = zion, df = TRUE)
```

Such results can be used to generate summary statistics for raster values per polygon, for example to characterize a single region or to compare many regions. The generation of summary statistics is demonstrated in the code below, which creates the object `zion_srtm_df` containing summary statistics for elevation values in Zion National Park (see Figure 5.20(A)):

```
group_by(zion_srtm_values, ID) %>%
  summarize_at(vars(srtm), list(~min, ~mean, ~max))
#> # A tibble: 1 x 4
#>      ID   min  mean   max
#>   <dbl> <dbl> <dbl> <dbl>
#> 1     1  1122 1818.  2661
```

The preceding code chunk used the **tidyverse** to provide summary statistics for cell values per polygon ID, as described in Chapter 3. The results provide useful summaries, for example that the maximum height in the park is around 2,661 meters (other summary statistics, such as standard deviation, can also be calculated in this way). Because there is only one polygon in the example a data frame with a single row is returned; however, the method works when multiple selector polygons are used.

The same approach works for counting occurrences of categorical raster values within polygons. This is illustrated with a land cover dataset (`nlcd`) from the **spDataLarge** package in Figure 5.20(B), and demonstrated in the code below:

```
zion_nlcd = raster::extract(nlcd, zion, df = TRUE, factors = TRUE)
dplyr::select(zion_nlcd, ID, levels) %>%
  tidyr::gather(key, value, -ID) %>%
  group_by(ID, key, value) %>%
  tally() %>%
  tidyr::spread(value, n, fill = 0)
#> # A tibble: 1 x 9
#> # Groups:   ID, key [1]
#>      ID key    Barren Cultivated Developed  Forest Herbaceous Shrubland
#>   <dbl> <chr>   <dbl>      <dbl>     <dbl>   <dbl>      <dbl>     <dbl>
#> 1     1 leve~   98285         62      4205  298299        235    203701
#> # ... with 1 more variable: Wetlands <dbl>
```

So far, we have seen how `raster::extract()` is a flexible way of extracting raster cell values from a range of input geographic objects. An issue with the function, however, is that it is relatively slow. If this is a problem, it is useful to know about alternatives and work-arounds, three of which are presented below.

- **Parallelization**: this approach works when using many geographic vector selector objects by splitting them into groups and extracting cell values independently for each group (see `?raster::clusterR()` for details of this approach).
- Use the **velox** package (Hunziker, 2017), which provides a fast method for

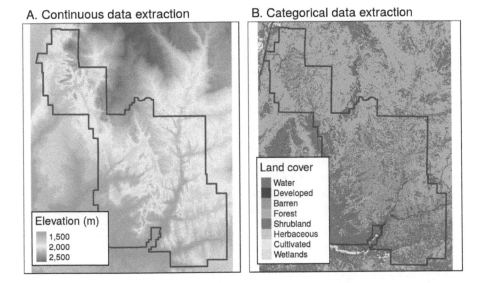

FIGURE 5.20: Area used for continuous (left) and categorical (right) raster extraction.

extracting raster data that fits in memory (see the packages `extract`[7] vignette for details).

- Using **R-GIS bridges** (see Chapter 9): efficient calculation of raster statistics from polygons can be found in the SAGA function `saga:gridstatisticsforpolygons`, for example, which can be accessed via **RQGIS**.

5.4.3 Rasterization

Rasterization is the conversion of vector objects into their representation in raster objects. Usually, the output raster is used for quantitative analysis (e.g., analysis of terrain) or modeling. As we saw in Chapter 2 the raster data model has some characteristics that make it conducive to certain methods. Furthermore, the process of rasterization can help simplify datasets because the resulting values all have the same spatial resolution: rasterization can be seen as a special type of geographic data aggregation.

The **raster** package contains the function `rasterize()` for doing this work. Its first two arguments are, x, vector object to be rasterized and, y, a 'template raster' object defining the extent, resolution and CRS of the output. The geographic resolution of the input raster has a major impact on the results: if it is too low (cell size is too large), the result may miss the full geographic

---

[7]https://hunzikp.github.io/velox/extract.html

variability of the vector data; if it is too high, computational times may be excessive. There are no simple rules to follow when deciding an appropriate geographic resolution, which is heavily dependent on the intended use of the results. Often the target resolution is imposed on the user, for example when the output of rasterization needs to be aligned to the existing raster.

To demonstrate rasterization in action, we will use a template raster that has the same extent and CRS as the input vector data `cycle_hire_osm_projected` (a dataset on cycle hire points in London is illustrated in Figure 5.21(A)) and spatial resolution of 1000 meters:

```
cycle_hire_osm_projected = st_transform(cycle_hire_osm, 27700)
raster_template = raster(extent(cycle_hire_osm_projected), resolution = 1000,
                    crs = st_crs(cycle_hire_osm_projected)$proj4string)
```

Rasterization is a very flexible operation: the results depend not only on the nature of the template raster, but also on the type of input vector (e.g., points, polygons) and a variety of arguments taken by the `rasterize()` function.

To illustrate this flexibility we will try three different approaches to rasterization. First, we create a raster representing the presence or absence of cycle hire points (known as presence/absence rasters). In this case `rasterize()` requires only one argument in addition to x and y (the aforementioned vector and raster objects): a value to be transferred to all non-empty cells specified by `field` (results illustrated Figure 5.21(B)).

```
ch_raster1 = rasterize(cycle_hire_osm_projected, raster_template, field = 1)
```

The `fun` argument specifies summary statistics used to convert multiple observations in close proximity into associate cells in the raster object. By default `fun = "last"` is used but other options such as `fun = "count"` can be used, in this case to count the number of cycle hire points in each grid cell (the results of this operation are illustrated in Figure 5.21(C)).

```
ch_raster2 = rasterize(cycle_hire_osm_projected, raster_template,
                    field = 1, fun = "count")
```

The new output, `ch_raster2`, shows the number of cycle hire points in each grid cell. The cycle hire locations have different numbers of bicycles described by the `capacity` variable, raising the question, what's the capacity in each grid cell? To calculate that we must `sum` the field ("capacity"), resulting in output illustrated in Figure 5.21(D), calculated with the following command (other summary functions such as `mean` could be used):

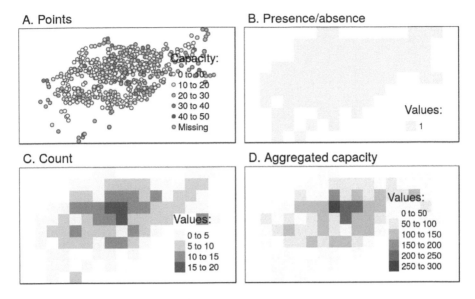

FIGURE 5.21: Examples of point rasterization.

```
ch_raster3 = rasterize(cycle_hire_osm_projected, raster_template,
                field = "capacity", fun = sum)
```

Another dataset based on California's polygons and borders (created below) illustrates rasterization of lines. After casting the polygon objects into a multilinestring, a template raster is created with a resolution of a 0.5 degree:

```
california = dplyr::filter(us_states, NAME == "California")
california_borders = st_cast(california, "MULTILINESTRING")
raster_template2 = raster(extent(california), resolution = 0.5,
                crs = st_crs(california)$proj4string)
```

Line rasterization is demonstrated in the code below. In the resulting raster, all cells that are touched by a line get a value, as illustrated in Figure 5.22(A).

```
california_raster1 = rasterize(california_borders, raster_template2)
```

Polygon rasterization, by contrast, selects only cells whose centroids are inside the selector polygon, as illustrated in Figure 5.22(B).

```
california_raster2 = rasterize(california, raster_template2)
```

As with `raster::extract()`, `raster::rasterize()` works well for most cases but

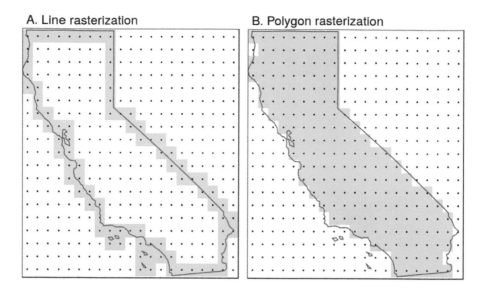

FIGURE 5.22: Examples of line and polygon rasterizations.

is not performance optimized. Fortunately, there are several alternatives, including the `fasterize::fasterize()` and `gdalUtils::gdal_rasterize()`. The former is much (100 times+) faster than `rasterize()`, but is currently limited to polygon rasterization. The latter is part of GDAL and therefore requires a vector file (instead of an `sf` object) and rasterization parameters (instead of a `Raster*` template object) as inputs.[8]

5.4.4   Spatial vectorization

Spatial vectorization is the counterpart of rasterization (Section 5.4.3), but in the opposite direction. It involves converting spatially continuous raster data into spatially discrete vector data such as points, lines or polygons.

> Be careful with the wording! In R, vectorization refers to the possibility of replacing `for`-loops and alike by doing things like `1:10 / 2` (see also Wickham (2014a)).

The simplest form of vectorization is to convert the centroids of raster cells into points. `rasterToPoints()` does exactly this for all non-`NA` raster grid cells (Figure 5.23). Setting the `spatial` parameter to `TRUE` ensures the output is a spatial object, not a matrix.

---

[8]See more at `http://gdal.org/gdal_rasterize.html`.

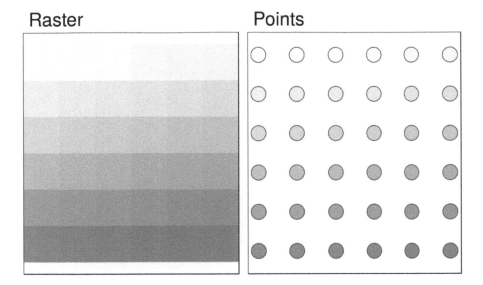

FIGURE 5.23: Raster and point representation of the elev object.

```
elev_point = rasterToPoints(elev, spatial = TRUE) %>%
  st_as_sf()
```

Another common type of spatial vectorization is the creation of contour lines representing lines of continuous height or temperatures (isotherms) for example. We will use a real-world digital elevation model (DEM) because the artificial raster elev produces parallel lines (task: verify this and explain why this happens). Contour lines can be created with the **raster** function rasterToContour(), which is itself a wrapper around contourLines(), as demonstrated below (not shown):

```
data(dem, package = "RQGIS")
cl = rasterToContour(dem)
plot(dem, axes = FALSE)
plot(cl, add = TRUE)
```

Contours can also be added to existing plots with functions such as contour(), rasterVis::contourplot() or tmap::tm_iso(). As illustrated in Figure 5.24, isolines can be labelled.

```
# create hillshade
hs = hillShade(slope = terrain(dem, "slope"), aspect = terrain(dem, "aspect"))
plot(hs, col = gray(0:100 / 100), legend = FALSE)
```

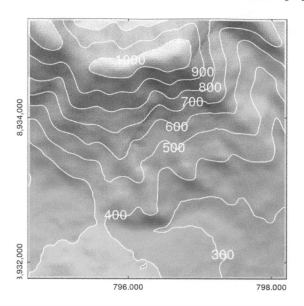

FIGURE 5.24: DEM hillshade of the southern flank of Mt. Mongón overlaid by contour lines.

```
# overlay with DEM
plot(dem, col = terrain.colors(25), alpha = 0.5, legend = FALSE, add = TRUE)
# add contour lines
contour(dem, col = "white", add = TRUE)
```

The final type of vectorization involves conversion of rasters to polygons. This can be done with raster::rasterToPolygons(), which converts each raster cell into a polygon consisting of five coordinates, all of which are stored in memory (explaining why rasters are often fast compared with vectors!).

This is illustrated below by converting the grain object into polygons and subsequently dissolving borders between polygons with the same attribute values (also see the dissolve argument in rasterToPolygons()). Attributes in this case are stored in a column called layer (see Section 5.2.6 and Figure 5.25). (Note: a convenient alternative for converting rasters into polygons is spex::polygonize() which by default returns an sf object.)

```
grain_poly = rasterToPolygons(grain) %>%
  st_as_sf()
grain_poly2 = grain_poly %>%
```

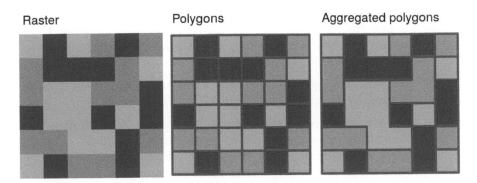

FIGURE 5.25: Illustration of vectorization of raster (left) into polygon (center) and polygon aggregation (right).

```
group_by(layer) %>%
summarize()
```

5.5   Exercises

Some of the exercises use a vector (`random_points`) and raster dataset (`ndvi`) from the **RQGIS** package. They also use a polygonal 'convex hull' derived from the vector dataset (`ch`) to represent the area of interest:

```
library(RQGIS)
data(random_points)
data(ndvi)
ch = st_combine(random_points) %>%
  st_convex_hull()
```

1. Generate and plot simplified versions of the `nz` dataset. Experiment with different values of `keep` (ranging from 0.5 to 0.00005) for `ms_simplify()` and `dTolerance` (from 100 to 100,000) `st_simplify()` .

   - At what value does the form of the result start to break down for each method, making New Zealand unrecognizable?
   - Advanced: What is different about the geometry type of the results from `st_simplify()` compared with the geometry type of `ms_simplify()`? What problems does this create and how can this be resolved?

2. In the first exercise in Chapter 4 it was established that Canterbury region had 70 of the 101 highest points in New Zealand. Using `st_buffer()`, how many points in `nz_height` are within 100 km of Canterbury?

3. Find the geographic centroid of New Zealand. How far is it from the geographic centroid of Canterbury?

4. Most world maps have a north-up orientation. A world map with a south-up orientation could be created by a reflection (one of the affine transformations not mentioned in Section 5.2.4) of the `world` object's geometry. Write code to do so. Hint: you need to use a two-element vector for this transformation.

   • Bonus: create an upside-down map of your country.

5. Subset the point in `p` that is contained within `x` *and* `y` (see Section 5.2.5 and Figure 5.8).

   • Using base subsetting operators.
   • Using an intermediary object created with `st_intersection()`.

6. Calculate the length of the boundary lines of US states in meters. Which state has the longest border and which has the shortest? Hint: The `st_length` function computes the length of a LINESTRING or MULTILINESTRING geometry.

7. Crop the `ndvi` raster using (1) the `random_points` dataset and (2) the `ch` dataset. Are there any differences in the output maps? Next, mask `ndvi` using these two datasets. Can you see any difference now? How can you explain that?

8. Firstly, extract values from `ndvi` at the points represented in `random_points`. Next, extract average values of `ndvi` using a 90 buffer around each point from `random_points` and compare these two sets of values. When would extracting values by buffers be more suitable than by points alone?

9. Subset points higher than 3100 meters in New Zealand (the `nz_height` object) and create a template raster with a resolution of 3 km. Using these objects:

   • Count numbers of the highest points in each grid cell.
   • Find the maximum elevation in each grid cell.

10. Aggregate the raster counting high points in New Zealand (created in the previous exercise), reduce its geographic resolution by half (so cells are 6 by 6 km) and plot the result.

    • Resample the lower resolution raster back to a resolution of 3 km. How have the results changed?

- Name two advantages and disadvantages of reducing raster resolution.

11. Polygonize the `grain` dataset and filter all squares representing clay.

- Name two advantages and disadvantages of vector data over raster data.
- At which points would it be useful to convert rasters to vectors in your work?

6

Reprojecting geographic data

Prerequisites

- This chapter requires the following packages (**lwgeom** is also used, but does not need to be attached):

```
library(sf)
library(raster)
library(dplyr)
library(spData)
library(spDataLarge)
```

6.1 Introduction

Section 2.4 introduced coordinate reference systems (CRSs) and demonstrated their importance. This chapter goes further. It highlights issues that can arise when using inappropriate CRSs and how to *transform* data from one CRS to another.

As illustrated in Figure 2.1, there are two types of CRSs: *geographic* ('lon/lat', with units in degrees longitude and latitude) and *projected* (typically with units of meters from a datum). This has consequences. Many geometry operations in **sf**, for example, assume their inputs have a projected CRS, because the GEOS functions they are based on assume projected data. To deal with this issue **sf** provides the function `st_is_longlat()` to check. In some cases the CRS is unknown, as shown below using the example of London introduced in Section 2.2:

```
london = data.frame(lon = -0.1, lat = 51.5) %>%
  st_as_sf(coords = c("lon", "lat"))
```

```
st_is_longlat(london)
#> [1] NA
```

This shows that unless a CRS is manually specified or is loaded from a source that has CRS metadata, the CRS is NA. A CRS can be added to sf objects with st_set_crs() as follows:[1]

```
london_geo = st_set_crs(london, 4326)
st_is_longlat(london_geo)
#> [1] TRUE
```

Datasets without a specified CRS can cause problems. An example is provided below, which creates a buffer of one unit around london and london_geo objects:

```
london_buff_no_crs = st_buffer(london, dist = 1)
london_buff = st_buffer(london_geo, dist = 1)
#> Warning in st_buffer.sfc(st_geometry(x), dist, nQuadSegs, endCapStyle =
#> endCapStyle, : st_buffer does not correctly buffer longitude/latitude data
#> dist is assumed to be in decimal degrees (arc_degrees).
```

Only the second operation generates a warning. The warning message is useful, telling us that the result may be of limited use because it is in units of latitude and longitude, rather than meters or some other suitable measure of distance assumed by st_buffer(). The consequences of a failure to work on projected data are illustrated in Figure 6.1 (left panel): the buffer is elongated in the north-south direction because lines of longitude converge towards the Earth's poles.

> The distance between two lines of longitude, called meridians, is around 111 km at the equator (execute geosphere::distGeo(c(0, 0), c(1, 0)) to find the precise distance). This shrinks to zero at the poles. At the latitude of London, for example, meridians are less than 70 km apart (challenge: execute code that verifies this). Lines of latitude, by contrast, have constant distance from each other irrespective of latitude: they are always around 111 km apart, including at the equator and near the poles. This is illustrated in Figures 6.1 and 6.3.

Do not interpret the warning about the geographic (longitude/latitude) CRS as "the CRS should not be set": it almost always should be! It is better understood as a suggestion to *reproject* the data onto a projected CRS. This suggestion

---

[1]The CRS can also be added when creating sf objects with the crs argument (e.g., st_sf(geometry = st_sfc(st_point(c(-0.1, 51.5))), crs = 4326)). The same argument can also be used to set the CRS when creating raster datasets (e.g., raster(crs = "+proj=longlat")).

does not always need to be heeded: performing spatial and geometric operations makes little or no difference in some cases (e.g., spatial subsetting). But for operations involving distances such as buffering, the only way to ensure a good result is to create a projected copy of the data and run the operation on that. This is done in the code chunk below:

```
london_proj = data.frame(x = 530000, y = 180000) %>%
  st_as_sf(coords = 1:2, crs = 27700)
```

The result is a new object that is identical to `london`, but reprojected onto a suitable CRS (the British National Grid, which has an EPSG code of 27700 in this case) that has units of meters. We can verify that the CRS has changed using `st_crs()` as follows (some of the output has been replaced by ...):

```
st_crs(london_proj)
#> Coordinate Reference System:
#>    EPSG: 27700
#>    proj4string: "+proj=tmerc +lat_0=49 +lon_0=-2 ... +units=m +no_defs"
```

Notable components of this CRS description include the EPSG code (`EPSG: 27700`), the projection (transverse Mercator[2], `+proj=tmerc`), the origin (`+lat_0=49 +lon_0=-2`) and units (`+units=m`).[3] The fact that the units of the CRS are meters (rather than degrees) tells us that this is a projected CRS: `st_is_longlat(london_proj)` now returns `FALSE` and geometry operations on `london_proj` will work without a warning, meaning buffers can be produced from it using proper units of distance. As pointed out above, moving one degree means moving a bit more than 111 km at the equator (to be precise: 111,320 meters). This is used as the new buffer distance:

```
london_proj_buff = st_buffer(london_proj, 111320)
```

The result in Figure 6.1 (right panel) shows that buffers based on a projected CRS are not distorted: every part of the buffer's border is equidistant to London.

The importance of CRSs (primarily whether they are projected or geographic) has been demonstrated using the example of London. The subsequent sections go into more depth, exploring which CRS to use and the details of reprojecting vector and raster objects.

---

[2]https://en.wikipedia.org/wiki/Transverse_Mercator_projection

[3]For a short description of the most relevant projection parameters and related concepts, see the fourth lecture by Jochen Albrecht hosted at http://www.geography.hunter.cuny.edu/~jochen/GTECH361/lectures/ and information at https://proj4.org/parameters.html. Other great resources on projections are spatialreference.org and progonos.com/furuti/MapProj.

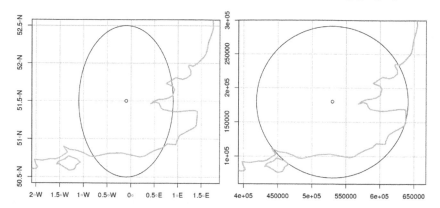

FIGURE 6.1: Buffers around London with a geographic (left) and projected (right) CRS. The gray outline represents the UK coastline.

6.2   When to reproject?

The previous section showed how to set the CRS manually, with `st_set_crs(london, 4326)`. In real world applications, however, CRSs are usually set automatically when data is read-in. The main task involving CRSs is often to *transform* objects, from one CRS into another. But when should data be transformed? And into which CRS? There are no clear-cut answers to these questions and CRS selection always involves trade-offs (Maling, 1992). However, there are some general principles provided in this section that can help you decide.

First it's worth considering *when to transform*. In some cases transformation to a projected CRS is essential, such as when using geometric functions such as `st_buffer()`, as Figure 6.1 shows. Conversely, publishing data online with the **leaflet** package may require a geographic CRS. Another case is when two objects with different CRSs must be compared or combined, as shown when we try to find the distance between two objects with different CRSs:

```
st_distance(london_geo, london_proj)
# > Error: st_crs(x) == st_crs(y) is not TRUE
```

To make the `london` and `london_proj` objects geographically comparable one of them must be transformed into the CRS of the other. But which CRS to use? The answer is usually 'to the projected CRS', which in this case is the British National Grid (EPSG:27700):

```
london2 = st_transform(london_geo, 27700)
```

Now that a transformed version of london has been created, using the **sf** function
st_transform(), the distance between the two representations of London can be
found. It may come as a surprise that london and london2 are just over 2 km
apart![4]

```
st_distance(london2, london_proj)
#> Units: [m]
#>      [,1]
#> [1,] 2018
```

6.3   Which CRS to use?

The question of *which CRS* is tricky, and there is rarely a 'right' answer:
"There exist no all-purpose projections, all involve distortion when far from the
center of the specified frame" (Bivand et al., 2013). For geographic CRSs, the
answer is often WGS84[5], not only for web mapping (covered in the previous
paragraph) but also because GPS datasets and thousands of raster and vector
datasets are provided in this CRS by default. WGS84 is the most common
CRS in the world, so it is worth knowing its EPSG code: 4326. This 'magic
number' can be used to convert objects with unusual projected CRSs into
something that is widely understood.

What about when a projected CRS is required? In some cases, it is not
something that we are free to decide: "often the choice of projection is made
by a public mapping agency" (Bivand et al., 2013). This means that when
working with local data sources, it is likely preferable to work with the CRS in
which the data was provided, to ensure compatibility, even if the official CRS
is not the most accurate. The example of London was easy to answer because
(a) the British National Grid (with its associated EPSG code 27700) is well
known and (b) the original dataset (london) already had that CRS.

In cases where an appropriate CRS is not immediately clear, the choice of
CRS should depend on the properties that are most important to preserve in

---

[4]The difference in location between the two points is not due to imperfections in the
transforming operation (which is in fact very accurate) but the low precision of the manually-
created coordinates that created london and london_proj. Also surprising may be that the
result is provided in a matrix with units of meters. This is because st_distance() can provide
distances between many features and because the CRS has units of meters. Use as.numeric()
to coerce the result into a regular number.

[5]https://en.wikipedia.org/wiki/World_Geodetic_System#A_new_World_Geodetic_System:_WGS_84

the subsequent maps and analysis. All CRSs are either equal-area, equidistant, conformal (with shapes remaining unchanged), or some combination of compromises of those. Custom CRSs with local parameters can be created for a region of interest and multiple CRSs can be used in projects when no single CRS suits all tasks. 'Geodesic calculations' can provide a fall-back if no CRSs are appropriate (see proj4.org/geodesic.html[6]). For any projected CRS the results may not be accurate when used on geometries covering hundreds of kilometers.

When deciding a custom CRS, we recommend the following:[7]

- A Lambert azimuthal equal-area (LAEA[8]) projection for a custom local projection (set lon_0 and lat_0 to the center of the study area), which is an equal-area projection at all locations but distorts shapes beyond thousands of kilometres.

- Azimuthal equidistant (AEQD[9]) projections for a specifically accurate straight-line distance between a point and the centre point of the local projection.

- Lambert conformal conic (LCC[10]) projections for regions covering thousands of kilometres, with the cone set to keep distance and area properties reasonable between the secant lines.

- Stereographic (STERE[11]) projections for polar regions, but taking care not to rely on area and distance calculations thousands of kilometres from the center.

A commonly used default is Universal Transverse Mercator (UTM[12]), a set of CRSs that divides the Earth into 60 longitudinal wedges and 20 latitudinal segments. The transverse Mercator projection used by UTM CRSs is conformal but distorts areas and distances with increasing severity with distance from the center of the UTM zone. Documentation from the GIS software Manifold therefore suggests restricting the longitudinal extent of projects using UTM zones to 6 degrees from the central meridian (source: manifold.net[13]).

Almost every place on Earth has a UTM code, such as "60H" which refers to northern New Zealand where R was invented. All UTM projections have the same datum (WGS84) and their EPSG codes run sequentially from 32601 to 32660 (for northern hemisphere locations) and 32701 to 32760 (southern hemisphere locations).

---

[6]https://proj4.org/geodesic.html
[7]Many thanks to an anonymous reviewer whose comments formed the basis of this advice.
[8]https://en.wikipedia.org/wiki/Lambert_azimuthal_equal-area_projection
[9]https://en.wikipedia.org/wiki/Azimuthal_equidistant_projection
[10]https://en.wikipedia.org/wiki/Lambert_conformal_conic_projection
[11]https://en.wikipedia.org/wiki/Stereographic_projection
[12]https://en.wikipedia.org/wiki/Universal_Transverse_Mercator_coordinate_system
[13]http://www.manifold.net/doc/mfd9/universal_transverse_mercator_projection.htm

To show how the system works, let's create a function, lonlat2UTM() to calculate the EPSG code associated with any point on the planet as follows[14]:

```
lonlat2UTM = function(lonlat) {
  utm = (floor((lonlat[1] + 180) / 6) %% 60) + 1
  if(lonlat[2] > 0) {
    utm + 32600
  } else{
    utm + 32700
  }
}
```

The following command uses this function to identify the UTM zone and associated EPSG code for Auckland and London:

```
epsg_utm_auk = lonlat2UTM(c(174.7, -36.9))
epsg_utm_lnd = lonlat2UTM(st_coordinates(london))
st_crs(epsg_utm_auk)$proj4string
#> [1] "+proj=utm +zone=60 +south +datum=WGS84 +units=m +no_defs"
st_crs(epsg_utm_lnd)$proj4string
#> [1] "+proj=utm +zone=30 +datum=WGS84 +units=m +no_defs"
```

Maps of UTM zones such as that provided by dmap.co.uk[15] confirm that London is in UTM zone 30U.

Another approach to automatically selecting a projected CRS specific to a local dataset is to create an azimuthal equidistant (AEQD[16]) projection for the center-point of the study area. This involves creating a custom CRS (with no EPSG code) with units of meters based on the centerpoint of a dataset. This approach should be used with caution: no other datasets will be compatible with the custom CRS created and results may not be accurate when used on extensive datasets covering hundreds of kilometers.

Although we used vector datasets to illustrate the points outlined in this section, the principles apply equally to raster datasets. The subsequent sections explain features of CRS transformation that are unique to each geographic data model, continuing with vector data in the next section (Section 6.4) and moving on to explain how raster transformation is different, in Section 6.6.

---

[14] https://stackoverflow.com/a/9188972/
[15] http://www.dmap.co.uk/utmworld.htm
[16] https://en.wikipedia.org/wiki/Azimuthal_equidistant_projection

6.4   Reprojecting vector geometries

Chapter 2 demonstrated how vector geometries are made-up of points, and
how points form the basis of more complex objects such as lines and polygons.
Reprojecting vectors thus consists of transforming the coordinates of these
points. This is illustrated by `cycle_hire_osm`, an `sf` object from **spData** that
represents cycle hire locations across London. The previous section showed
how the CRS of vector data can be queried with `st_crs()`. Although the output
of this function is printed as a single entity, the result is in fact a named list
of class `crs`, with names `proj4string` (which contains full details of the CRS)
and `epsg` for its code. This is demonstrated below:

```
crs_lnd = st_crs(cycle_hire_osm)
class(crs_lnd)
#> [1] "crs"
crs_lnd$epsg
#> [1] 4326
```

This duality of CRS objects means that they can be set either using an
EPSG code or a `proj4string`. This means that `st_crs("+proj=longlat +datum=WGS84
+no_defs")` is equivalent to `st_crs(4326)`, although not all `proj4strings` have an
associated EPSG code. Both elements of the CRS are changed by transforming
the object to a projected CRS:

```
cycle_hire_osm_projected = st_transform(cycle_hire_osm, 27700)
```

The resulting object has a new CRS with an EPSG code 27700. But how to
find out more details about this EPSG code, or any code? One option is to
search for it online. Another option is to use a function from the **rgdal** package
to find the name of the CRS:

```
crs_codes = rgdal::make_EPSG()[1:2]
dplyr::filter(crs_codes, code == 27700)
#>    code                                   note
#> 1 27700 # OSGB 1936 / British National Grid
```

The result shows that the EPSG code 27700 represents the British National
Grid, a result that could have been found by searching online for "EPSG
27700[17]". But what about the `proj4string` element? `proj4strings` are text strings
in a particular format the describe the CRS. They can be seen as formulas for

---
[17]`https://www.google.com/search?q=CRS+27700`

converting a projected point into a point on the surface of the Earth and can be accessed from `crs` objects as follows (see proj4.org[18] for further details of what the output means):

```
st_crs(27700)$proj4string
#> [1] "+proj=tmerc +lat_0=49 +lon_0=-2 +k=0.9996012717 +x_0=400000 ...
```

Printing a spatial object in the console, automatically returns its coordinate reference system. To access and modify it explicitly, use the `st_crs` function, for example, `st_crs(cycle_hire_osm)`.

6.5 Modifying map projections

Established CRSs captured by EPSG codes are well-suited for many applications. However in some cases it is desirable to create a new CRS, using a custom `proj4string`. This system allows a very wide range of projections to be created, as we'll see in some of the custom map projections in this section.

A long and growing list of projections has been developed and many of these these can be set with the `+proj=` element of `proj4strings`.[19] When mapping the world while preserving area relationships, the Mollweide projection is a good choice (Jenny et al., 2017) (Figure 6.2). To use this projection, we need to specify it using the `proj4string` element, `"+proj=moll"`, in the `st_transform` function:

```
world_mollweide = st_transform(world, crs = "+proj=moll")
```

On the other hand, when mapping the world, it is often desirable to have as little distortion as possible for all spatial properties (area, direction, distance). One of the most popular projections to achieve this is the Winkel tripel projection (Figure 6.3).[20] `st_transform_proj()` from the **lwgeom** package which allows for coordinate transformations to a wide range of CRSs, including the Winkel tripel projection:

```
world_wintri = lwgeom::st_transform_proj(world, crs = "+proj=wintri")
```

---

[18] http://proj4.org/
[19] The Wikipedia page 'List of map projections' has 70+ projections and illustrations.
[20] This projection is used, among others, by the National Geographic Society.

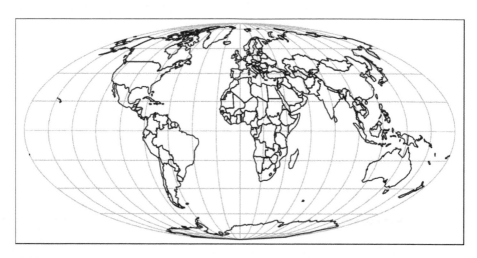

FIGURE 6.2: Mollweide projection of the world.

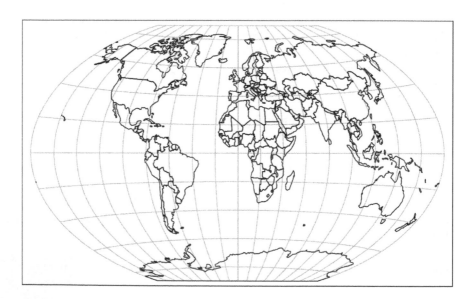

FIGURE 6.3: Winkel tripel projection of the world.

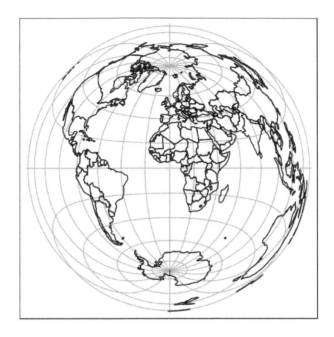

FIGURE 6.4: Lambert azimuthal equal-area projection of the world centered on longitude and latitude of 0.

The three main functions for transformation of simple features coordinates are `sf::st_transform()`, `sf::sf_project()`, and `lwgeom::st_transform_proj()`. The `st_transform` function uses the GDAL interface to PROJ, while `sf_project()` (which works with two-column numeric matrices, representing points) and `lwgeom::st_transform_proj()` use the PROJ API directly. The first one is appropriate for most situations, and provides a set of the most often used parameters and well-defined transformations. The next one allows for greater customization of a projection, which includes cases when some of the PROJ parameters (e.g., `+over`) or projection (`+proj=wintri`) is not available in `st_transform()`.

Moreover, PROJ parameters can be modified in most CRS definitions. The below code transforms the coordinates to the Lambert azimuthal equal-area projection centered on longitude and latitude of 0 (Figure 6.4).

```
world_laea1 = st_transform(world,
                    crs = "+proj=laea +x_0=0 +y_0=0 +lon_0=0 +lat_0=0")
```

We can change the PROJ parameters, for example the center of the projection, using the `+lon_0` and `+lat_0` parameters. The code below gives the map centered on New York City (Figure 6.5).

FIGURE 6.5: Lambert azimuthal equal-area projection of the world centered on New York City.

```
world_laea2 = st_transform(world,
                           crs = "+proj=laea +x_0=0 +y_0=0 +lon_0=-74 +lat_0=40")
```

More information on CRS modifications can be found in the Using PROJ[21] documentation.

6.6   Reprojecting raster geometries

The projection concepts described in the previous section apply equally to rasters. However, there are important differences in reprojection of vectors and rasters: transforming a vector object involves changing the coordinates of every vertex but this does not apply to raster data. Rasters are composed of rectangular cells of the same size (expressed by map units, such as degrees or meters), so it is impossible to transform coordinates of pixels separately. Raster reprojection involves creating a new raster object, often with a different number

---

[21]http://proj4.org/usage/index.html

of columns and rows than the original. The attributes must subsequently be re-estimated, allowing the new pixels to be 'filled' with appropriate values. In other words, raster reprojection can be thought of as two separate spatial operations: a vector reprojection of cell centroids to another CRS (Section 6.4), and computation of new pixel values through resampling (Section 5.3.3). Thus in most cases when both raster and vector data are used, it is better to avoid reprojecting rasters and reproject vectors instead.

The raster reprojection process is done with `projectRaster()` from the **raster** package. Like the `st_transform()` function demonstrated in the previous section, `projectRaster()` takes a geographic object (a raster dataset in this case) and a `crs` argument. However, `projectRaster()` only accepts the lengthy `proj4string` definitions of a CRS rather than concise EPSG codes.

It is possible to use a EPSG code in a `proj4string` definition with `"+init=epsg:MY_NUMBER"`. For example, one can use the `"+init=epsg:4326"` definition to set CRS to WGS84 (EPSG code of 4326). The PROJ library automatically adds the rest of the parameters and converts them into `"+init=epsg:4326 +proj=longlat +datum=WGS84 +no_defs +ellps=WGS84 +towgs84=0,0,0"`.

Let's take a look at two examples of raster transformation: using categorical and continuous data. Land cover data are usually represented by categorical maps. The `nlcd2011.tif` file provides information for a small area in Utah, USA obtained from National Land Cover Database 2011[22] in the NAD83 / UTM zone 12N CRS.

```
cat_raster = raster(system.file("raster/nlcd2011.tif", package = "spDataLarge"))
crs(cat_raster)
#> CRS arguments:
#>  +proj=utm +zone=12 +ellps=GRS80 +towgs84=0,0,0,0,0,0,0 +units=m
#> +no_defs
```

In this region, 14 land cover classes were distinguished (a full list of NLCD2011 land cover classes can be found at mrlc.gov[23]):

```
unique(cat_raster)
#>  [1] 11 21 22 23 31 41 42 43 52 71 81 82 90 95
```

When reprojecting categorical rasters, the estimated values must be the same as those of the original. This could be done using the nearest neighbor method (`ngb`). This method assigns new cell values to the nearest cell center of the input raster. An example is reprojecting `cat_raster` to WGS84, a geographic CRS

---

[22]https://www.mrlc.gov/nlcd2011.php
[23]https://www.mrlc.gov/nlcd11_leg.php

TABLE 6.1: Key attributes in the original and projected categorical raster datasets.

CRS	nrow	ncol	ncell	resolution	unique_categories
NAD83	1359	1073	1458207	31.5275	14
WGS84	1394	1111	1548734	0.0003	15

well suited for web mapping. The first step is to obtain the PROJ definition of this CRS, which can be done using the http://spatialreference.org[24] webpage. The final step is to reproject the raster with the projectRaster() function which, in the case of categorical data, uses the nearest neighbor method (ngb):

```
wgs84 = "+proj=longlat +ellps=WGS84 +datum=WGS84 +no_defs"
cat_raster_wgs84 = projectRaster(cat_raster, crs = wgs84, method = "ngb")
```

Many properties of the new object differ from the previous one, including the number of columns and rows (and therefore number of cells), resolution (transformed from meters into degrees), and extent, as illustrated in Table 6.1 (note that the number of categories increases from 14 to 15 because of the addition of NA values, not because a new category has been created — the land cover classes are preserved).

Reprojecting numeric rasters (with numeric or in this case integer values) follows an almost identical procedure. This is demonstrated below with srtm.tif in **spDataLarge** from the Shuttle Radar Topography Mission (SRTM)[25], which represents height in meters above sea level (elevation) with the WGS84 CRS:

```
con_raster = raster(system.file("raster/srtm.tif", package = "spDataLarge"))
crs(con_raster)
#> CRS arguments:
#>  +proj=longlat +datum=WGS84 +no_defs +ellps=WGS84 +towgs84=0,0,0
```

We will reproject this dataset into a projected CRS, but *not* with the nearest neighbor method which is appropriate for categorical data. Instead, we will use the bilinear method which computes the output cell value based on the four nearest cells in the original raster. The values in the projected dataset are the distance-weighted average of the values from these four cells: the closer the input cell is to the center of the output cell, the greater its weight. The following commands create a text string representing the Oblique Lambert

---

[24]http://spatialreference.org/ref/epsg/wgs-84/
[25]https://www2.jpl.nasa.gov/srtm/

TABLE 6.2: Key attributes original and projected continuous raster datasets.

CRS	nrow	ncol	ncell	resolution	mean
WGS84	457	465	212505	31.5275	1843
Equal-area	467	478	223226	0.0003	1842

azimuthal equal-area projection, and reproject the raster into this CRS, using the `bilinear` method:

```
equalarea = "+proj=laea +lat_0=37.32 +lon_0=-113.04"
con_raster_ea = projectRaster(con_raster, crs = equalarea, method = "bilinear")
crs(con_raster_ea)
#> CRS arguments:
#>  +proj=laea +lat_0=37.32 +lon_0=-113.04 +ellps=WGS84
```

Raster reprojection on numeric variables also leads to small changes to values and spatial properties, such as the number of cells, resolution, and extent. These changes are demonstrated in Table 6.2[26]:

> Of course, the limitations of 2D Earth projections apply as much to vector as to raster data. At best we can comply with two out of three spatial properties (distance, area, direction). Therefore, the task at hand determines which projection to choose. For instance, if we are interested in a density (points per grid cell or inhabitants per grid cell) we should use an equal-area projection (see also Chapter 13).

There is more to learn about CRSs. An excellent resource in this area, also implemented in R, is the website R Spatial. Chapter 6 for this free online book is recommended reading — see: rspatial.org/spatial/rst/6-crs.html[27]

6.7 Exercises

1. Create a new object called `nz_wgs` by transforming `nz` object into the WGS84 CRS.

---

[26]Another minor change, that is not represented in Table 6.2, is that the class of the values in the new projected raster dataset is `numeric`. This is because the `bilinear` method works with continuous data and the results are rarely coerced into whole integer values. This can have implications for file sizes when raster datasets are saved.

[27]http://rspatial.org/spatial/rst/6-crs.html

- Create an object of class `crs` for both and use this to query their CRSs.
- With reference to the bounding box of each object, what units does each CRS use?
- Remove the CRS from `nz_wgs` and plot the result: what is wrong with this map of New Zealand and why?

2. Transform the `world` dataset to the transverse Mercator projection (`"+proj=tmerc"`) and plot the result. What has changed and why? Try to transform it back into WGS 84 and plot the new object. Why does the new object differ from the original one?

3. Transform the continuous raster (`cat_raster`) into WGS 84 using the nearest neighbor interpolation method. What has changed? How does it influence the results?

4. Transform the categorical raster (`cat_raster`) into WGS 84 using the bilinear interpolation method. What has changed? How does it influence the results?

5. Create your own `proj4string`. It should have the Lambert Azimuthal Equal Area (`laea`) projection, the WGS84 ellipsoid, the longitude of projection center of 95 degrees west, the latitude of projection center of 60 degrees north, and its units should be in meters. Next, subset Canada from the `world` object and transform it into the new projection. Plot and compare a map before and after the transformation.

7

Geographic data I/O

Prerequisites

This chapter requires the following packages:

```
library(sf)
library(raster)
library(dplyr)
library(spData)
```

7.1 Introduction

This chapter is about reading and writing geographic data. Geographic data *import* is essential for geocomputation: real-world applications are impossible without data. For others to benefit from the results of your work, data *output* is also vital. Taken together, we refer to these processes as I/O, short for input/output.

Geographic data I/O is almost always part of a wider process. It depends on knowing which datasets are *available*, where they can be *found* and how to *retrieve* them. These topics are covered in Section 7.2, which describes various *geoportals*, which collectively contain many terabytes of data, and how to use them. To further ease data access, a number of packages for downloading geographic data have been developed. These are described in Section 7.3.

There are many geographic file formats, each of which has pros and cons. These are described in Section 7.5. The process of actually reading and writing such file formats efficiently is not covered until Sections 7.6 and 7.7, respectively. The final Section 7.8 demonstrates methods for saving visual outputs (maps), in preparation for Chapter 8 on visualization.

7.2   Retrieving open data

A vast and ever-increasing amount of geographic data is available on the
internet, much of which is free to access and use (with appropriate credit given
to its providers). In some ways there is now *too much* data, in the sense that
there are often multiple places to access the same dataset. Some datasets are
of poor quality. In this context, it is vital to know where to look, so the first
section covers some of the most important sources. Various 'geoportals' (web
services providing geospatial datasets such as Data.gov[1]) are a good place to
start, providing a wide range of data but often only for specific locations (as
illustrated in the updated Wikipedia page[2] on the topic).

Some global geoportals overcome this issue. The GEOSS portal[3] and the
Copernicus Open Access Hub[4], for example, contain many raster datasets with
global coverage. A wealth of vector datasets can be accessed from the National
Aeronautics and Space Administration agency (NASA), SEDAC[5] portal and
the European Union's INSPIRE geoportal[6], with global and regional coverage.

Most geoportals provide a graphical interface allowing datasets to be queried
based on characteristics such spatial and temporal extent, the United States
Geological Services' EarthExplorer[7] being a prime example. *Exploring* datasets
interactively on a browser is an effective way of understanding available layers.
Downloading data is best done with code, however, from reproducibility and
efficiency perspectives. Downloads can be initiated from the command line
using a variety of techniques, primarily via URLs and APIs (see the Sentinel
API[8] for example). Files hosted on static URLs can be downloaded with
`download.file()`, as illustrated in the code chunk below which accesses US
National Parks data from: catalog.data.gov/dataset/national-parks[9]:

```
download.file(url = "http://nrdata.nps.gov/programs/lands/nps_boundary.zip",
              destfile = "nps_boundary.zip")
unzip(zipfile = "nps_boundary.zip")
usa_parks = st_read(dsn = "nps_boundary.shp")
```

---

[1] https://catalog.data.gov/dataset?metadata_type=geospatial

[2] https://en.wikipedia.org/wiki/Geoportal

[3] http://www.geoportal.org/

[4] https://scihub.copernicus.eu/

[5] http://sedac.ciesin.columbia.edu/

[6] http://inspire-geoportal.ec.europa.eu/

[7] https://earthexplorer.usgs.gov/

[8] https://scihub.copernicus.eu/twiki/do/view/SciHubWebPortal/APIHubDescription

[9] https://catalog.data.gov/dataset/national-parks

TABLE 7.1: Selected R packages for geographic data retrieval.

Package	Description
getlandsat	Provides access to Landsat 8 data.
osmdata	Download and import of OpenStreetMap data.
raster	getData() imports administrative, elevation, WorldClim data.
rnaturalearth	Access to Natural Earth vector and raster data.
rnoaa	Imports National Oceanic and Atmospheric Administration (NOAA) climate data.
rWBclimate	Access World Bank climate data.

7.3 Geographic data packages

A multitude of R packages have been developed for accessing geographic data, some of which are presented in Table 7.1. These provide interfaces to one or more spatial libraries or geoportals and aim to make data access even quicker from the command line.

It should be emphasised that Table 7.1 represents only a small number of available geographic data packages. Other notable packages include **GSODR**, which provides Global Summary Daily Weather Data in R (see the package's README[10] for an overview of weather data sources); **tidycensus** and **tigris**, which provide socio-demographic vector data for the USA; and **hddtools**, which provides access to a range of hydrological datasets.

Each data package has its own syntax for accessing data. This diversity is demonstrated in the subsequent code chunks, which show how to get data using three packages from Table 7.1. Country borders are often useful and these can be accessed with the `ne_countries()` function from the **rnaturalearth** package as follows:

```
library(rnaturalearth)
usa = ne_countries(country = "United States of America") # United States borders
class(usa)
#> [1] "SpatialPolygonsDataFrame"
#> attr(,"package")
#> [1] "sp"
# alternative way of accessing the data, with raster::getData()
# getData("GADM", country = "USA", level = 0)
```

By default **rnaturalearth** returns objects of class `spatial`. The result can be converted into an `sf` objects with `st_as_sf()` as follows:

---

[10] https://github.com/ropensci/GSODR

```
usa_sf = st_as_sf(usa)
```

A second example downloads a series of rasters containing global monthly
precipitation sums with spatial resolution of ten minutes. The result is a
multilayer object of class `RasterStack`.

```
library(raster)
worldclim_prec = getData(name = "worldclim", var = "prec", res = 10)
class(worldclim_prec)
#> [1] "RasterStack"
#> attr(,"package")
#> [1] "raster"
```

A third example uses the **osmdata** package (Padgham et al., 2018) to find parks
from the OpenStreetMap (OSM) database. As illustrated in the code-chunk
below, queries begin with the function `opq()` (short for OpenStreetMap query),
the first argument of which is bounding box, or text string representing a
bounding box (the city of Leeds in this case). The result is passed to a
function for selecting which OSM elements we're interested in (parks in this
case), represented by *key-value pairs*. Next, they are passed to the function
`osmdata_sf()` which does the work of downloading the data and converting it
into a list of `sf` objects (see `vignette('osmdata')` for further details):

```
library(osmdata)
parks = opq(bbox = "leeds uk") %>%
  add_osm_feature(key = "leisure", value = "park") %>%
  osmdata_sf()
```

OpenStreetMap is a vast global database of crowd-sourced data and it is
growing daily. Although the quality is not as spatially consistent as many
official datasets, OSM data have many advantages: they are globally available
free of charge and using crowd-source data can encourage 'citizen science' and
contributions back to the digital commons. Further examples of **osmdata** in
action are provided in Chapters 9, 12 and 13.

Sometimes, packages come with inbuilt datasets. These can be accessed in four
ways: by attaching the package (if the package uses 'lazy loading' as **spData**
does), with `data(dataset)`, by referring to the dataset with `pkg::dataset` or with
`system.file()` to access raw data files. The following code chunk illustrates
the latter two options using the `world` (already loaded by attaching its parent
package with `library(spData)`):[11]

---

[11]For more information on data import with R packages, see Sections 5.5 and 5.6 of
Gillespie and Lovelace (2016).

```
world2 = spData::world
world3 = st_read(system.file("shapes/world.gpkg", package = "spData"))
```

7.4 Geographic web services

In an effort to standardize web APIs for accessing spatial data, the Open Geospatial Consortium (OGC) has created a number of specifications for web services (collectively known as OWS, which is short for OGC Web Services). These specifications include the Web Feature Service (WFS), Web Map Service (WMS), Web Map Tile Service (WMTS), the Web Coverage Service (WCS) and even a Wep Processing Service (WPS). Map servers such as PostGIS have adopted these protocols, leading to standardization of queries: Like other web APIs, OWS APIs use a 'base URL' and a 'query string' proceding a ? to request data.

There are many requests that can be made to a OWS service. One of the most fundamental is getCapabilities, demonstrated with the **httr** package to show how API queries can be constructed and dispatched, in this case to discover the capabilities of a service providing run by the Food and Agriculture Organization of the United Nations (FAO):

```
base_url = "http://www.fao.org/figis/geoserver/wfs"
q = list(request = "GetCapabilities")
res = httr::GET(url = base_url, query = q)
res$url
#> [1] "http://www.fao.org/figis/geoserver/wfs?request=GetCapabilities"
```

The above code chunk demonstrates how API requests can be constructed programmatically with the GET() function, which takes a base URL and a list of query parameters which can easily be extended. The result of the request is saved in res, an object of class response defined in the **httr** package, which is a list containing information of the request, including the URL. As can be seen by executing browseURL(res$url), the results can also be read directly in a browser. One way of extracting the contents of the request is as follows:

```
txt = httr::content(res, "text")
xml = xml2::read_xml(txt)
```

```
#> {xml_document} ...
```

```
#> [1] <ows:ServiceIdentification>\n  <ows:Title>GeoServer WFS...
#> [2] <ows:ServiceProvider>\n  <ows:ProviderName>Food and Agr...
#> ...
```

Data can be downloaded from WFS services with the `GetFeature` request and a specific `typeName` (as illustrated in the code chunk below).

Available names differ depending on the accessed web feature service. One can extract them programmatically using web technologies (Nolan and Lang, 2014) or scrolling manually through the contents of the `GetCapabilities` output in a browser.

```
qf = list(request = "GetFeature", typeName = "area:FAO_AREAS")
file = tempfile(fileext = ".gml")
httr::GET(url = base_url, query = qf, httr::write_disk(file))
fao_areas = sf::read_sf(file)
```

Note the use of `write_disc()` to ensure that the results are written to disk rather than loaded into memory, allowing them to be imported with **sf**. This example shows how to gain low-level access to web services using **httr**, which can be useful for understanding how web services work. For many everyday tasks, however, a higher-level interface may be more appropriate, and a number of R packages, and tutorials, have been developed precisely for this purpose.

Packages **ows4R**, **rwfs** and **sos4R** have been developed for working with OWS services in general, WFS and the sensor observation service (SOS) respectively. As of October 2018, only **ows4R** is on CRAN. The package's basic functionality is demonstrated below, in commands that get all `FAO_AREAS` as we did in the previous code chunk:[12]

```
library(ows4R)
wfs = WFSClient$new("http://www.fao.org/figis/geoserver/wfs",
                    serviceVersion = "1.0.0", logger = "INFO")
fao_areas = wfs$getFeatures("area:FAO_AREAS")
```

There is much more to learn about web services and much potential for development of R-OWS interfaces, an active area of development. For further information on the topic, we recommend examples from European Centre for Medium-Range Weather Forecasts (ECMWF) services at

---

[12]To filter features on the server before downloading them, the argument `cql_filter` can be used. Adding `cql_filter = URLencode("F_CODE= '27'")` to the command, for example, would instruct the server to only return the feature with values in the `F_CODE` column equal to 27.

github.com/OpenDataHack[13] and reading-up on OCG Web Services at opengeospatial.org[14].

7.5 File formats

Geographic datasets are usually stored as files or in spatial databases. File formats can either store vector or raster data, while spatial databases such as PostGIS[15] can store both (see also Section 9.6.2). Today the variety of file formats may seem bewildering but there has been much consolidation and standardization since the beginnings of GIS software in the 1960s when the first widely distributed program (SYMAP[16]) for spatial analysis was created at Harvard University (Coppock and Rhind, 1991).

GDAL (which should be pronounced "goo-dal", with the double "o" making a reference to object-orientation), the Geospatial Data Abstraction Library, has resolved many issues associated with incompatibility between geographic file formats since its release in 2000. GDAL provides a unified and high-performance interface for reading and writing of many raster and vector data formats. Many open and proprietary GIS programs, including GRASS, ArcGIS and QGIS, use GDAL behind their GUIs for doing the legwork of ingesting and spitting out geographic data in appropriate formats.

GDAL provides access to more than 200 vector and raster data formats. Table 7.2 presents some basic information about selected and often used spatial file formats.

An important development ensuring the standardization and open-sourcing of file formats was the founding of the Open Geospatial Consortium (OGC[17]) in 1994. Beyond defining the simple features data model (see Section 2.2.1), the OGC also coordinates the development of open standards, for example as used in file formats such as KML and GeoPackage. Open file formats of the kind endorsed by the OGC have several advantages over proprietary formats: the standards are published, ensure transparency and open up the possibility for users to further develop and adjust the file formats to their specific needs.

ESRI Shapefile is the most popular vector data exchange format. However, it is not an open format (though its specification is open). It was developed in the early 1990s and has a number of limitations. First of all, it is a multi-file format, which consists of at least three files. It only supports 255 columns,

---

[13]https://github.com/OpenDataHack/data_service_catalogue

[14]http://www.opengeospatial.org/standards

[15]https://trac.osgeo.org/postgis/wiki/WKTRaster

[16]https://news.harvard.edu/gazette/story/2011/10/the-invention-of-gis/

[17]http://www.opengeospatial.org/

TABLE 7.2: Selected spatial file formats.

Name	Extension	Comments
ESRI Shapefile	.shp	Popular vector format consisting of at least three files. No support for: files > 2GB; mixed types; names > 10 chars; cols > 255.
GeoJSON	.geojson	Open vector format, extends JSON by including a subset of the simple feature representation.
KML	.kml	XML-based open vector format developed for use with Google Earth. Zipped KML file forms the KMZ format.
GPX	.gpx	Open XML schema created for exchange of vector GPS data.
GeoTIFF	.tiff	Popular open raster format similar to '.tif' format but stores raster header.
Arc ASCII	.asc	Open raster format where the first six lines represent the raster header, followed by the raster cell values arranged in rows and columns.
R-raster	.gri, .grd	Open raster format of the R-package raster.
SQLite/SpatiaLite	.sqlite	Standalone open relational database, SpatiaLite extends SQLite to support raster and vector data.
ESRI FileGDB	.gdb	Proprietary format for raster and vector data. Allows: multiple feature classes; topology. Limited support from GDAL.
GeoPackage	.gpkg	Lightweight open format based on SQLite allowing an easy and platform-independent exchange of vector and raster data.

column names are restricted to ten characters and the file size limit is 2 GB. Furthermore, ESRI Shapefile does not support all possible geometry types, for example, it is unable to distinguish between a polygon and a multipolygon.[18] Despite these limitations, a viable alternative had been missing for a long time. In the meantime, GeoPackage[19] emerged, and seems to be a more than suitable replacement candidate for ESRI Shapefile. Geopackage is a format for exchanging geospatial information and an OGC standard. The GeoPackage standard describes the rules on how to store geospatial information in a tiny SQLite container. Hence, GeoPackage is a lightweight spatial database container, which allows the storage of vector and raster data but also of

---

[18]To learn more about ESRI Shapefile limitations and possible alternative file formats, visit http://switchfromshapefile.org/.
[19]https://www.geopackage.org/

TABLE 7.3: Sample of available drivers for reading/writing vector data (it could vary between different GDAL versions).

name	long_name	write	is_raster	is_vector
ESRI Shapefile	ESRI Shapefile	TRUE	FALSE	TRUE
GPX	GPX	TRUE	FALSE	TRUE
KML	Keyhole Markup Language (KML)	TRUE	FALSE	TRUE
GeoJSON	GeoJSON	TRUE	FALSE	TRUE
GPKG	GeoPackage	TRUE	TRUE	TRUE

non-spatial data and extensions. Aside from GeoPackage, there are other geospatial data exchange formats worth checking out (Table 7.2).

7.6 Data input (I)

Executing commands such as `sf::st_read()` (the main function we use for loading vector data) or `raster::raster()` (the main function used for loading raster data) silently sets off a chain of events that reads data from files. Moreover, there are many R packages containing a wide range of geographic data or providing simple access to different data sources. All of them load the data into R or, more precisely, assign objects to your workspace, stored in RAM accessible from the `.GlobalEnv`[20] of the R session.

7.6.1 Vector data

Spatial vector data comes in a wide variety of file formats, most of which can be read-in via the **sf** function `st_read()`. Behind the scenes this calls GDAL. To find out which data formats **sf** supports, run `st_drivers()`. Here, we show only the first five drivers (see Table 7.3):

```
sf_drivers = st_drivers()
head(sf_drivers, n = 5)
```

The first argument of `st_read()` is `dsn`, which should be a text string or an object containing a single text string. The content of a text string could vary

---

[20]http://adv-r.had.co.nz/Environments.html

between different drivers. In most cases, as with the ESRI Shapefile (.shp) or the GeoPackage format (.gpkg), the dsn would be a file name. st_read() guesses the driver based on the file extension, as illustrated for a .gpkg file below:

```
vector_filepath = system.file("shapes/world.gpkg", package = "spData")
world = st_read(vector_filepath)
#> Reading layer 'world' from data source '.../world.gpkg' using driver 'GPKG'
#> Simple feature collection with 177 features and 10 fields
#> geometry type:  MULTIPOLYGON
#> dimension:      XY
#> bbox:           xmin: -180 ymin: -90 xmax: 180 ymax: 83.64513
#> epsg (SRID):    4326
#> proj4string:    +proj=longlat +datum=WGS84 +no_defs
```

For some drivers, dsn could be provided as a folder name, access credentials for a database, or a GeoJSON string representation (see the examples of the st_read() help page for more details).

Some vector driver formats can store multiple data layers. By default, st_read() automatically reads the first layer of the file specified in dsn; however, using the layer argument you can specify any other layer.

Naturally, some options are specific to certain drivers.[21] For example, think of coordinates stored in a spreadsheet format (.csv). To read in such files as spatial objects, we naturally have to specify the names of the columns (x and y in our example below) representing the coordinates. We can do this with the help of the options parameter. To find out about possible options, please refer to the 'Open Options' section of the corresponding GDAL driver description. For the comma-separated value (csv) format, visit http://www.gdal.org/drv_csv.html.

```
cycle_hire_txt = system.file("misc/cycle_hire_xy.csv", package = "spData")
cycle_hire_xy = st_read(cycle_hire_txt, options = c("X_POSSIBLE_NAMES=X",
                                                    "Y_POSSIBLE_NAMES=Y"))
```

Instead of columns describing xy-coordinates, a single column can also contain the geometry information. Well-known text (WKT), well-known binary (WKB), and the GeoJSON formats are examples of this. For instance, the world_wkt.csv file has a column named WKT representing polygons of the world's countries. We will again use the options parameter to indicate this. Here, we will use read_sf() which does exactly the same as st_read() except it does not print the driver name to the console and stores strings as characters instead of factors.

---

[21] A list of supported vector formats and options can be found at http://gdal.org/ogr_formats.html.

```
world_txt = system.file("misc/world_wkt.csv", package = "spData")
world_wkt = read_sf(world_txt, options = "GEOM_POSSIBLE_NAMES=WKT")
# the same as
world_wkt = st_read(world_txt, options = "GEOM_POSSIBLE_NAMES=WKT",
                    quiet = TRUE, stringsAsFactors = FALSE)
```

Not all of the supported vector file formats store information about their coordinate reference system. In these situations, it is possible to add the missing information using the st_set_crs() function. Please refer also to Section 2.4 for more information.

As a final example, we will show how st_read() also reads KML files. A KML file stores geographic information in XML format - a data format for the creation of web pages and the transfer of data in an application-independent way (Nolan and Lang, 2014). Here, we access a KML file from the web. This file contains more than one layer. st_layers() lists all available layers. We choose the first layer Placemarks and say so with the help of the layer parameter in read_sf().

```
u = "https://developers.google.com/kml/documentation/KML_Samples.kml"
download.file(u, "KML_Samples.kml")
st_layers("KML_Samples.kml")
#> Driver: LIBKML
#> Available layers:
#>                 layer_name geometry_type features fields
#> 1                Placemarks                      3     11
#> 2         Styles and Markup                      1     11
#> 3          Highlighted Icon                      1     11
#> 4            Ground Overlays                     1     11
#> 5            Screen Overlays                     0     11
#> 6                     Paths                      6     11
#> 7                  Polygons                      0     11
#> 8             Google Campus                      4     11
#> 9           Extruded Polygon                     1     11
#> 10 Absolute and Relative                        4     11
kml = read_sf("KML_Samples.kml", layer = "Placemarks")
```

All the examples presented in this section so far have used the **sf** package for geographic data import. It is fast and flexible but it may be worth looking at other packages for specific file formats. An example is the **geojsonsf** package.

A benchmark[22] suggests it is around 10 times faster than the **sf** package for reading `.geojson`.

7.6.2   Raster data

Similar to vector data, raster data comes in many file formats with some of them supporting even multilayer files. **raster**'s `raster()` command reads in a single layer.

```
raster_filepath = system.file("raster/srtm.tif", package = "spDataLarge")
single_layer = raster(raster_filepath)
```

In case you want to read in a single band from a multilayer file, use the `band` parameter to indicate a specific layer.

```
multilayer_filepath = system.file("raster/landsat.tif", package = "spDataLarge")
band3 = raster(multilayer_filepath, band = 3)
```

If you want to read in all bands, use `brick()` or `stack()`.

```
multilayer_brick = brick(multilayer_filepath)
multilayer_stack = stack(multilayer_filepath)
```

Please refer to Section 2.3.3 for information on the difference between raster stacks and bricks.

7.7   Data output (O)

Writing geographic data allows you to convert from one format to another and to save newly created objects. Depending on the data type (vector or raster), object class (e.g., `multipoint` or `RasterLayer`), and type and amount of stored information (e.g., object size, range of values), it is important to know how to store spatial files in the most efficient way. The next two sections will demonstrate how to do this.

7.7.1   Vector data

The counterpart of `st_read()` is `st_write()`. It allows you to write **sf** objects to a wide range of geographic vector file formats, including the most common

---

[22]`https://github.com/ATFutures/geobench`

such as `.geojson`, `.shp` and `.gpkg`. Based on the file name, `st_write()` decides automatically which driver to use. The speed of the writing process depends also on the driver.

```
st_write(obj = world, dsn = "world.gpkg")
#> Writing layer 'world' to data source 'world.gpkg' using driver 'GPKG'
#> features:       177
#> fields:         10
#> geometry type:  Multi Polygon
```

Note: if you try to write to the same data source again, the function will fail:

```
st_write(obj = world, dsn = "world.gpkg")
#> Updating layer 'world' to data source 'world.gpkg' using driver 'GPKG'
#> Creating layer world failed.
#> Error in CPL_write_ogr(obj, dsn, layer, driver, ...),  :
#>   Layer creation failed.
#> In addition: Warning message:
#> In CPL_write_ogr(obj, dsn, layer, driver, ...),  :
#>   GDAL Error 1: Layer world already exists, CreateLayer failed.
#> Use the layer creation option OVERWRITE=YES to replace it.
```

The error message provides some information as to why the function failed. The GDAL Error 1 statement makes clear that the failure occurred at the GDAL level. Additionally, the suggestion to use OVERWRITE=YES provides a clue about how to fix the problem. However, this is not a `st_write()` argument, it is a GDAL option. Luckily, `st_write` provides a `layer_options` argument through which we can pass driver-dependent options:

```
st_write(obj = world, dsn = "world.gpkg", layer_options = "OVERWRITE=YES")
```

Another solution is to use the `st_write()` argument `delete_layer`. Setting it to TRUE deletes already existing layers in the data source before the function attempts to write (note there is also a `delete_dsn` argument):

```
st_write(obj = world, dsn = "world.gpkg", delete_layer = TRUE)
```

You can achieve the same with `write_sf()` since it is equivalent to (technically an *alias* for) `st_write()`, except that its defaults for `delete_layer` and `quiet` is TRUE.

```
write_sf(obj = world, dsn = "world.gpkg")
```

The `layer_options` argument could be also used for many different purposes. One of them is to write spatial data to a text file. This can be done by specifying `GEOMETRY` inside of `layer_options`. It could be either `AS_XY` for simple point datasets (it creates two new columns for coordinates) or `AS_WKT` for more complex spatial data (one new column is created which contains the well-known text representation of spatial objects).

```
st_write(cycle_hire_xy, "cycle_hire_xy.csv", layer_options = "GEOMETRY=AS_XY")
st_write(world_wkt, "world_wkt.csv", layer_options = "GEOMETRY=AS_WKT")
```

7.7.2   Raster data

The `writeRaster()` function saves `Raster*` objects to files on disk. The function expects input regarding output data type and file format, but also accepts GDAL options specific to a selected file format (see `?writeRaster` for more details).

The **raster** package offers nine data types when saving a raster: LOG1S, INT1S, INT1U, INT2S, INT2U, INT4S, INT4U, FLT4S, and FLT8S.[23] The data type determines the bit representation of the raster object written to disk (Table 7.4). Which data type to use depends on the range of the values of your raster object. The more values a data type can represent, the larger the file will get on disk. Commonly, one would use LOG1S for bitmap (binary) rasters. Unsigned integers (INT1U, INT2U, INT4U) are suitable for categorical data, while float numbers (FLT4S and FLT8S) usually represent continuous data. `writeRaster()` uses FLT4S as the default. While this works in most cases, the size of the output file will be unnecessarily large if you save binary or categorical data. Therefore, we would recommend to use the data type that needs the least storage space, but is still able to represent all values (check the range of values with the `summary()` function).

The file extension determines the output file when saving a `Raster*` object to disk. For example, the `.tif` extension will create a GeoTIFF file:

```
writeRaster(x = single_layer,
            filename = "my_raster.tif",
            datatype = "INT2U")
```

The `raster` file format (native to the `raster` package) is used when a file extension is invalid or missing. Some raster file formats come with additional options. You can use them with the `options` parameter[24]. GeoTIFF files, for example, can be compressed using `COMPRESS`:

---

[23]Using INT4U is not recommended as R does not support 32-bit unsigned integers.
[24]http://www.gdal.org/formats_list.html

TABLE 7.4: Data types supported by the raster package.

Data type	Minimum value	Maximum value
LOG1S	FALSE (0)	TRUE (1)
INT1S	-127	127
INT1U	0	255
INT2S	-32,767	32,767
INT2U	0	65,534
INT4S	-2,147,483,647	2,147,483,647
INT4U	0	4,294,967,296
FLT4S	-3.4e+38	3.4e+38
FLT8S	-1.7e+308	1.7e+308

```
writeRaster(x = single_layer,
            filename = "my_raster.tif",
            datatype = "INT2U",
            options = c("COMPRESS=DEFLATE"),
            overwrite = TRUE)
```

Note that `writeFormats()` returns a list with all supported file formats on your computer.

7.8 Visual outputs

R supports many different static and interactive graphics formats. The most general method to save a static plot is to open a graphic device, create a plot, and close it, for example:

```
png(filename = "lifeExp.png", width = 500, height = 350)
plot(world["lifeExp"])
dev.off()
```

Other available graphic devices include `pdf()`, `bmp()`, `jpeg()`, `png()`, and `tiff()`. You can specify several properties of the output plot, including width, height and resolution.

Additionally, several graphic packages provide their own functions to save a graphical output. For example, the **tmap** package has the `tmap_save()` function. You can save a `tmap` object to different graphic formats by specifying the object name and a file path to a new graphic file.

```
library(tmap)
tmap_obj = tm_shape(world) +
  tm_polygons(col = "lifeExp")
tmap_save(tm  = tmap_obj, filename = "lifeExp_tmap.png")
```

On the other hand, you can save interactive maps created in the mapview package as an HTML file or image using the mapshot() function:

```
library(mapview)
mapview_obj = mapview(world, zcol = "lifeExp", legend = TRUE)
mapshot(mapview_obj, file = "my_interactive_map.html")
```

7.9 Exercises

1. List and describe three types of vector, raster, and geodatabase formats.

2. Name at least two differences between read_sf() and the more well-known function st_read().

3. Read the cycle_hire_xy.csv file from the **spData** package as a spatial object (Hint: it is located in the misc\ folder). What is a geometry type of the loaded object?

4. Download the borders of Germany using **rnaturalearth**, and create a new object called germany_borders. Write this new object to a file of the GeoPackage format.

5. Download the global monthly minimum temperature with a spatial resolution of five minutes using the **raster** package. Extract the June values, and save them to a file named tmin_june.tif file (hint: use raster::subset()).

6. Create a static map of Germany's borders, and save it to a PNG file.

7. Create an interactive map using data from the cycle_hire_xy.csv file. Export this map to a file called cycle_hire.html.

Part II

Extensions

8

Making maps with R

Prerequisites

- This chapter requires the following packages that we have already been using:

```
library(sf)
library(raster)
library(dplyr)
library(spData)
library(spDataLarge)
```

- In addition, it uses the following visualization packages:

```
library(tmap)     # for static and interactive maps
library(leaflet)  # for interactive maps
library(mapview)  # for interactive maps
library(ggplot2)  # tidyverse vis package
library(shiny)    # for web applications
```

8.1 Introduction

A satisfying and important aspect of geographic research is communicating the results. Map making — the art of cartography — is an ancient skill that involves communication, intuition, and an element of creativity. Static mapping is straightforward with `plot()`, as we saw in Section 2.2.3. It is possible to create advanced maps using base R methods (Murrell, 2016), but this chapter focuses on dedicated map-making packages. When learning a new skill, it makes sense to gain depth-of-knowledge in one area branching out. Map making is no exception, hence this chapter's coverage of one package (**tmap**) in depth rather than many superficially. In addition to being fun and creative, cartography also has important practical applications. A carefully crafted map is vital for effectively communicating the results of your work (Brewer, 2015):

Amateur-looking maps can undermine your audience's ability to understand important information and weaken the presentation of a professional data investigation.

Maps have been used for several thousand years for a wide variety of purposes. Historic examples include maps of buildings and land ownership in the Old Babylonian dynasty more than 3000 years ago and Ptolemy's world map in his masterpiece *Geography* nearly 2000 years ago (Talbert, 2014).

Map making has historically been an activity undertaken only by, or on behalf of, the elite. This has changed with the emergence of open source mapping software such as the R package **tmap** and the 'print composer' in QGIS which enable anyone to make high-quality maps, enabling 'citizen science'. Maps are also often the best way to present the findings of geocomputational research in a way that is accessible. Map making is therefore a critical part of geocomputation and its emphasis not only on describing, but also *changing* the world.

This chapter shows how to make a wide range of maps. The next section covers a range of static maps, including aesthetic considerations, facets and inset maps. Sections 8.3 to 8.5 cover animated and interactive maps (including web maps and mapping applications). Finally, Section 8.6 covers a range of alternative map-making packages including **ggplot2** and **cartogram**.

8.2   Static maps

Static maps are the most common type of visual output from geocomputation. Fixed images for printed outputs, common formats for static maps include .png and .pdf, for raster and vector outputs, respectively (interactive maps are covered in Section 8.4). Initially static maps were the *only* type of map that R could produce. Things have advanced greatly since **sp** was released (see Pebesma and Bivand, 2005). Many new techniques for map making have been developed since then. However, a decade later static plotting was still the emphasis of geographic data visualisation in R (Cheshire and Lovelace, 2015).

Despite the innovation of interactive mapping in R, static maps are still the foundation of mapping in R. The generic `plot()` function is often the fastest way to create static maps from vector and raster spatial objects, as shown in Sections 2.2.3 and 2.3.2. Sometimes simplicity and speed are priorities, especially during the development phase of a project, and this is where `plot()` excels. The base R approach is also extensible, with `plot()` offering dozens of arguments. Another low-level approach is the **grid** package, which provides functions for low-level control of graphical outputs, — see *R Graphics* (Murrell, 2016), especially Chapter 14[1]. The focus of this section, however, is making static maps with **tmap**.

Why **tmap**? It is a powerful and flexible map-making package with sensible defaults. It has a concise syntax that allows for the creation of attractive maps with minimal code, which will be familiar to **ggplot2** users. Furthermore, **tmap** has a unique capability to generate static and interactive maps using the same code via `tmap_mode()`. It accepts a wider range of spatial classes (including `raster` objects) than alternatives such as **ggplot2**, as documented in vignettes `tmap-getstarted`[2] and `tmap-changes-v2`[3] and an academic paper on the subject (Tennekes, 2018). This section teaches how to make static maps with **tmap**, emphasizing the important aesthetic and layout options.

8.2.1   tmap basics

Like **ggplot2**, **tmap** is based on the idea of a 'grammar of graphics' (Wilkinson and Wills, 2005). This involves a separation between the input data and the aesthetics (how data are visualised): each input dataset can be 'mapped' in a range of different ways including location on the map (defined by data's `geometry`), color, and other visual variables. The basic building block is `tm_shape()` (which defines input data, raster and vector objects), followed by one or more layer elements such as `tm_fill()` and `tm_dots()`. This layering is demonstrated in the chunk below, which generates the maps presented in Figure 8.1:

```
# Add fill layer to nz shape
tm_shape(nz) +
  tm_fill()
# Add border layer to nz shape
tm_shape(nz) +
  tm_borders()
# Add fill and border layers to nz shape
tm_shape(nz) +
```

---

[1]https://www.stat.auckland.ac.nz/~paul/RG2e/chapter14.html

[2]https://cran.r-project.org/web/packages/tmap/vignettes/tmap-getstarted.html

[3]https://cran.r-project.org/web/packages/tmap/vignettes/tmap-changes-v2.html

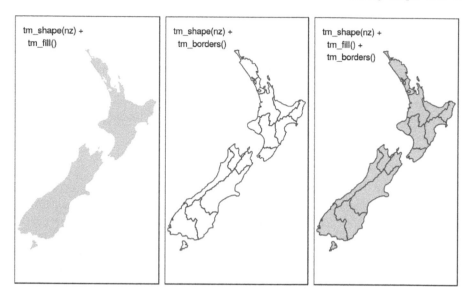

FIGURE 8.1: New Zealand's shape plotted with fill (left), border (middle) and fill and border (right) layers added using tmap functions.

```
tm_fill() +
tm_borders()
```

The object passed to `tm_shape()` in this case is `nz`, an `sf` object representing the regions of New Zealand (see Section 2.2.1 for more on `sf` objects). Layers are added to represent `nz` visually, with `tm_fill()` and `tm_borders()` creating shaded areas (left panel) and border outlines (middle panel) in Figure 8.1, respectively.

This is an intuitive approach to map making: the common task of *adding* new layers is undertaken by the addition operator `+`, followed by `tm_*()`. The asterisk (*) refers to a wide range of layer types which have self-explanatory names including `fill`, `borders` (demonstrated above), `bubbles`, `text` and `raster` (see `help("tmap-element")` for a full list). This layering is illustrated in the right panel of Figure 8.1, the result of adding a border *on top of* the fill layer.

`qtm()` is a handy function for **q**uickly creating **t**map **m**aps (hence the snappy name). It is concise and provides a good default visualization in many cases: `qtm(nz)`, for example, is equivalent to `tm_shape(nz) + tm_fill() + tm_borders()`. Further, layers can be added concisely using multiple `qtm()` calls, such as `qtm(nz) + qtm(nz_height)`. The disadvantage is that it makes aesthetics of individual layers harder to control, explaining why we avoid teaching it in this chapter.

8.2.2 Map objects

A useful feature of **tmap** is its ability to store *objects* representing maps. The code chunk below demonstrates this by saving the last plot in Figure 8.1 as an object of class `tmap` (note the use of `tm_polygons()` which condenses `tm_fill()` + `tm_borders()` into a single function):

```
map_nz = tm_shape(nz) + tm_polygons()
class(map_nz)
#> [1] "tmap"
```

`map_nz` can be plotted later, for example by adding additional layers (as shown below) or simply running `map_nz` in the console, which is equivalent to `print(map_nz)`.

New *shapes* can be added with `+ tm_shape(new_obj)`. In this case `new_obj` represents a new spatial object to be plotted on top of preceding layers. When a new shape is added in this way, all subsequent aesthetic functions refer to it, until another new shape is added. This syntax allows the creation of maps with multiple shapes and layers, as illustrated in the next code chunk which uses the function `tm_raster()` to plot a raster layer (with `alpha` set to make the layer semi-transparent):

```
map_nz1 = map_nz +
  tm_shape(nz_elev) + tm_raster(alpha = 0.7)
```

Building on the previously created `map_nz` object, the preceding code creates a new map object `map_nz1` that contains another shape (`nz_elev`) representing average elevation across New Zealand (see Figure 8.2, left). More shapes and layers can be added, as illustrated in the code chunk below which creates `nz_water`, representing New Zealand's territorial waters[4], and adds the resulting lines to an existing map object.

```
nz_water = st_union(nz) %>% st_buffer(22200) %>%
  st_cast(to = "LINESTRING")
map_nz2 = map_nz1 +
  tm_shape(nz_water) + tm_lines()
```

---

[4]https://en.wikipedia.org/wiki/Territorial_waters

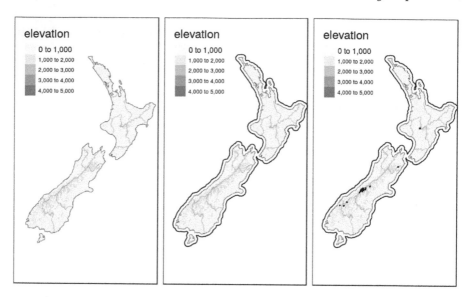

FIGURE 8.2: Maps with additional layers added to the final map of Figure 8.1.

There is no limit to the number of layers or shapes that can be added to `tmap` objects. The same shape can even be used multiple times. The final map illustrated in Figure 8.2 is created by adding a layer representing high points (stored in the object `nz_height`) onto the previously created `map_nz2` object with `tm_dots()` (see `?tm_dots` and `?tm_bubbles` for details on **tmap**'s point plotting functions). The resulting map, which has four layers, is illustrated in the right-hand panel of Figure 8.2:

```
map_nz3 = map_nz2 +
  tm_shape(nz_height) + tm_dots()
```

A useful and little known feature of **tmap** is that multiple map objects can be arranged in a single 'metaplot' with `tmap_arrange()`. This is demonstrated in the code chunk below which plots `map_nz1` to `map_nz3`, resulting in Figure 8.2.

```
tmap_arrange(map_nz1, map_nz2, map_nz3)
```

More elements can also be added with the + operator. Aesthetic settings, however, are controlled by arguments to layer functions.

8.2.3 Aesthetics

The plots in the previous section demonstrate **tmap**'s default aesthetic settings. Gray shades are used for `tm_fill()` and `tm_bubbles()` layers and a continuous black line is used to represent lines created with `tm_lines()`. Of course, these default values and other aesthetics can be overridden. The purpose of this section is to show how.

There are two main types of map aesthetics: those that change with the data and those that are constant. Unlike **ggplot2**, which uses the helper function `aes()` to represent variable aesthetics, **tmap** accepts aesthetic arguments that are either variable fields (based on column names) or constant values.[5] The most commonly used aesthetics for fill and border layers include color, transparency, line width and line type, set with `col`, `alpha`, `lwd`, and `lty` arguments, respectively. The impact of setting these with fixed values is illustrated in Figure 8.3.

```
ma1 = tm_shape(nz) + tm_fill(col = "red")
ma2 = tm_shape(nz) + tm_fill(col = "red", alpha = 0.3)
ma3 = tm_shape(nz) + tm_borders(col = "blue")
ma4 = tm_shape(nz) + tm_borders(lwd = 3)
ma5 = tm_shape(nz) + tm_borders(lty = 2)
ma6 = tm_shape(nz) + tm_fill(col = "red", alpha = 0.3) +
  tm_borders(col = "blue", lwd = 3, lty = 2)
tmap_arrange(ma1, ma2, ma3, ma4, ma5, ma6)
```

Like base R plots, arguments defining aesthetics can also receive values that vary. Unlike the base R code below (which generates the left panel in Figure 8.4), **tmap** aesthetic arguments will not accept a numeric vector:

```
plot(st_geometry(nz), col = nz$Land_area)  # works
tm_shape(nz) + tm_fill(col = nz$Land_area) # fails
#> Error: Fill argument neither colors nor valid variable name(s)
```

Instead `col` (and other aesthetics that can vary such as `lwd` for line layers and `size` for point layers) requires a character string naming an attribute associated with the geometry to be plotted. Thus, one would achieve the desired result as follows (plotted in the right-hand panel of Figure 8.4):

```
tm_shape(nz) + tm_fill(col = "Land_area")
```

An important argument in functions defining aesthetic layers such as `tm_fill()`

---

[5]If there is a clash between a fixed value and a column name, the column name takes precedence. This can be verified by running the next code chunk after running `nz$red = 1:nrow(nz)`.

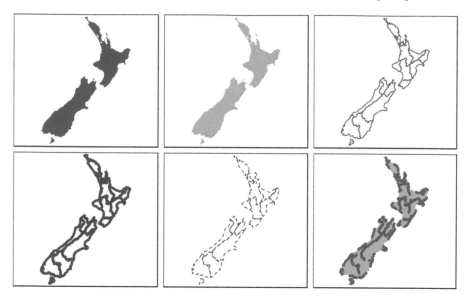

FIGURE 8.3: The impact of changing commonly used fill and border aesthetics to fixed values.

is title, which sets the title of the associated legend. The following code chunk demonstrates this functionality by providing a more attractive name than the variable name Land_area (note the use of expression() to create superscript text):

```
legend_title = expression("Area (km"^2*")")
map_nza = tm_shape(nz) +
  tm_fill(col = "Land_area", title = legend_title) + tm_borders()
```

8.2.4   Color settings

Color settings are an important part of map design. They can have a major impact on how spatial variability is portrayed as illustrated in Figure 8.5. This shows four ways of coloring regions in New Zealand depending on median income, from left to right (and demonstrated in the code chunk below):

- The default setting uses 'pretty' breaks, described in the next paragraph.
- breaks allows you to manually set the breaks.
- n sets the number of bins into which numeric variables are categorized.
- palette defines the color scheme, for example BuGn.

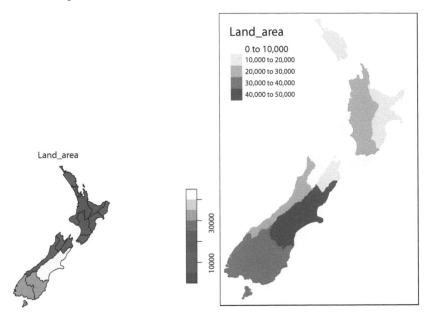

FIGURE 8.4: Comparison of base (left) and tmap (right) handling of a numeric color field.

```
tm_shape(nz) + tm_polygons(col = "Median_income")
breaks = c(0, 3, 4, 5) * 10000
tm_shape(nz) + tm_polygons(col = "Median_income", breaks = breaks)
tm_shape(nz) + tm_polygons(col = "Median_income", n = 10)
tm_shape(nz) + tm_polygons(col = "Median_income", palette = "BuGn")
```

Another way to change color settings is by altering color break (or bin) settings. In addition to manually setting `breaks` **tmap** allows users to specify algorithms to automatically create breaks with the `style` argument. Six of the most useful break styles are illustrated in Figure 8.6 and described in the bullet points below:

- `style = pretty`, the default setting, rounds breaks into whole numbers where possible and spaces them evenly.
- `style = equal` divides input values into bins of equal range, and is appropriate for variables with a uniform distribution (not recommended for variables with a skewed distribution as the resulting map may end-up having little color diversity).
- `style = quantile` ensures the same number of observations fall into each category (with the potential down side that bin ranges can vary widely).
- `style = jenks` identifies groups of similar values in the data and maximizes the differences between categories.

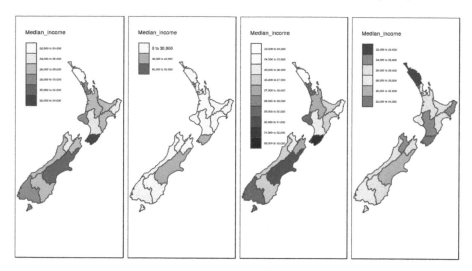

FIGURE 8.5: Illustration of settings that affect color settings. The results show (from left to right): default settings, manual breaks, n breaks, and the impact of changing the palette.

- `style = cont` (and `order`) present a large number of colors over continuous color field, and are particularly suited for continuous rasters (`order` can help visualize skewed distributions).
- `style = cat` was designed to represent categorical values and assures that each category receives a unique color.

Although `style` is an argument of **tmap** functions, in fact it originates as an argument in `classInt::classIntervals()` — see the help page of this function for details.

Palettes define the color ranges associated with the bins and determined by the `breaks`, `n`, and `style` arguments described above. The default color palette is specified in `tm_layout()` (see Section 8.2.5 to learn more); however, it could be quickly changed using the `palette` argument. It expects a vector of colors or a new color palette name, which can be selected interactively with `tmaptools::palette_explorer()`. You can add a - as prefix to reverse the palette order.

There are three main groups of color palettes: categorical, sequential and diverging (Figure 8.7), and each of them serves a different purpose. Categorical palettes consist of easily distinguishable colors and are most appropriate for categorical data without any particular order such as state names or land cover classes. Colors should be intuitive: rivers should be blue, for example, and

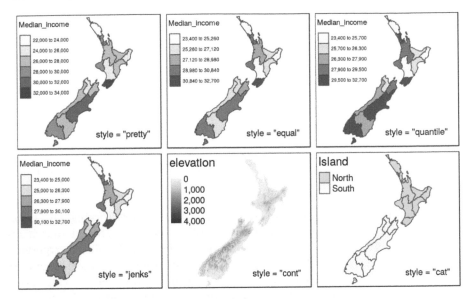

FIGURE 8.6: Illustration of different binning methods set using the style argument in tmap.

pastures green. Avoid too many categories: maps with large legends and many colors can be uninterpretable.[6]

The second group is sequential palettes. These follow a gradient, for example from light to dark colors (light colors tend to represent lower values), and are appropriate for continuous (numeric) variables. Sequential palettes can be single (Blues go from light to dark blue, for example) or multi-color/hue (YlOrBr is gradient from light yellow to brown via orange, for example), as demonstrated in the code chunk below — output not shown, run the code yourself to see the results!

```
tm_shape(nz) + tm_polygons("Population", palette = "Blues")
tm_shape(nz) + tm_polygons("Population", palette = "YlOrBr")
```

The last group, diverging palettes, typically range between three distinct colors (purple-white-green in Figure 8.7) and are usually created by joining two single-color sequential palettes with the darker colors at each end. Their main purpose is to visualize the difference from an important reference point, e.g., a certain temperature, the median household income or the mean probability for a drought event. The reference point's value can be adjusted in **tmap** using the midpoint argument.

---

[6]col = "MAP_COLORS" can be used in maps with a large number of individual polygons (for example, a map of individual countries) to create unique colors for adjacent polygons.

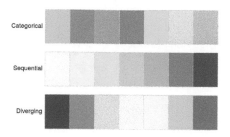

FIGURE 8.7: Examples of categorical, sequential and diverging palettes.

There are two important principles for consideration when working with colors: perceptibility and accessibility. Firstly, colors on maps should match our perception. This means that certain colors are viewed through our experience and also cultural lenses. For example, green colors usually represent vegetation or lowlands and blue is connected with water or cool. Color palettes should also be easy to understand to effectively convey information. It should be clear which values are lower and which are higher, and colors should change gradually. This property is not preserved in the rainbow color palette; therefore, we suggest avoiding it in geographic data visualization (Borland and Taylor II, 2007). Instead, the viridis color palettes[7], also available in **tmap**, can be used. Secondly, changes in colors should be accessible to the largest number of people. Therefore, it is important to use colorblind friendly palettes as often as possible.[8]

8.2.5   Layouts

The map layout refers to the combination of all map elements into a cohesive map. Map elements include among others the objects to be mapped, the title, the scale bar, margins and aspect ratios, while the color settings covered in the previous section relate to the palette and break-points used to affect how the map looks. Both may result in subtle changes that can have an equally large impact on the impression left by your maps.

Additional elements such as north arrows and scale bars have their own functions - `tm_compass()` and `tm_scale_bar()` (Figure 8.8).

```
map_nz +
  tm_compass(type = "8star", position = c("left", "top")) +
  tm_scale_bar(breaks = c(0, 100, 200), size = 1)
```

tmap also allows a wide variety of layout settings to be changed, some of

---

[7] https://cran.r-project.org/web/packages/viridis/

[8] See the "Color blindness simulator" options in `tmaptools::palette_explorer()`.

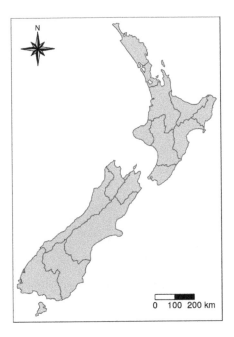

FIGURE 8.8: Map with additional elements - a north arrow and scale bar.

which are illustrated in Figure 8.9, produced using the following code (see `args(tm_layout)` or `?tm_layout` for a full list):

```
map_nz + tm_layout(title = "New Zealand")
map_nz + tm_layout(scale = 5)
map_nz + tm_layout(bg.color = "lightblue")
map_nz + tm_layout(frame = FALSE)
```

The other arguments in `tm_layout()` provide control over many more aspects of the map in relation to the canvas on which it is placed. Some useful layout settings are listed below (see Figure 8.10 for illustrations of a selection of these):

- Frame width (`frame.lwd`) and an option to allow double lines (`frame.double.line`).
- Margin settings including `outer.margin` and `inner.margin`.
- Font settings controlled by `fontface` and `fontfamily`.
- Legend settings including binary options such as `legend.show` (whether or not to show the legend) `legend.only` (omit the map) and `legend.outside` (should the legend go outside the map?), as well as multiple choice settings such as `legend.position`.

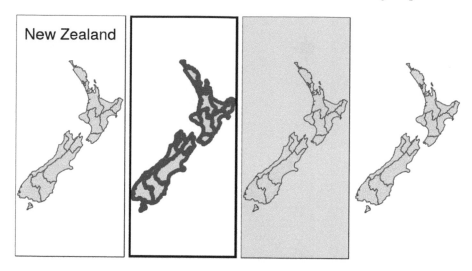

FIGURE 8.9: Layout options specified by (from left to right) title, scale, bg.color and frame arguments.

- Default colors of aesthetic layers (aes.color), map attributes such as the frame (attr.color).
- Color settings controlling sepia.intensity (how yellowy the map looks) and saturation (a color-grayscale).

The impact of changing the color settings listed above is illustrated in Figure 8.11 (see ?tm_layout for a full list).

Beyond the low-level control over layouts and colors, **tmap** also offers high-level styles, using the tm_style() function (representing the second meaning of 'style' in the package). Some styles such as tm_style("cobalt") result in stylized maps, while others such as tm_style("gray") make more subtle changes, as illustrated in Figure 8.12, created using code below (see 08-tmstyles.R):

```
map_nza + tm_style("bw")
map_nza + tm_style("classic")
map_nza + tm_style("cobalt")
map_nza + tm_style("col_blind")
```

A preview of predefined styles can be generated by executing tmap_style_catalogue(). This creates a folder called tmap_style_previews containing nine images. Each image, from tm_style_albatross.png to tm_style_white.png, shows a faceted map of the world in the corresponding style. Note: tmap_style_catalogue() takes some time to run.

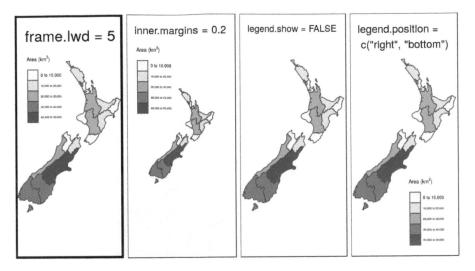

FIGURE 8.10: Illustration of selected layout options.

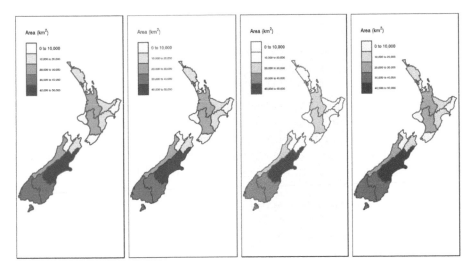

FIGURE 8.11: Illustration of selected color-related layout options.

8.2.6 Faceted maps

Faceted maps, also referred to as 'small multiples', are composed of many maps arranged side-by-side, and sometimes stacked vertically (Meulemans et al., 2017). Facets enable the visualization of how spatial relationships change with respect to another variable, such as time. The changing populations of settlements, for example, can be represented in a faceted map with each panel representing the population at a particular moment in time. The time

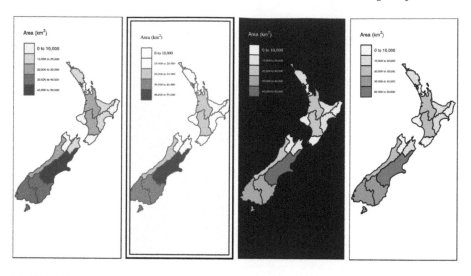

FIGURE 8.12: Selected tmap styles: bw, classic, cobalt and color blind (from left to right).

dimension could be represented via another *aesthetic* such as color. However, this risks cluttering the map because it will involve multiple overlapping points (cities do not tend to move over time!).

Typically all individual facets in a faceted map contain the same geometry data repeated multiple times, once for each column in the attribute data (this is the default plotting method for sf objects, see Chapter 2). However, facets can also represent shifting geometries such as the evolution of a point pattern over time. This use case of faceted plot is illustrated in Figure 8.13.

```
urb_1970_2030 = urban_agglomerations %>%
  filter(year %in% c(1970, 1990, 2010, 2030))
tm_shape(world) + tm_polygons() +
  tm_shape(urb_1970_2030) + tm_symbols(col = "black", border.col = "white",
                                       size = "population_millions") +
  tm_facets(by = "year", nrow = 2, free.coords = FALSE)
```

The preceding code chunk demonstrates key features of faceted maps created with **tmap**:

- Shapes that do not have a facet variable are repeated (the countries in world in this case).
- The by argument which varies depending on a variable (year in this case).
- nrow/ncol setting specifying the number of rows and columns that facets should be arranged into.
- The free.coords-parameter specifying if each map has its own bounding box.

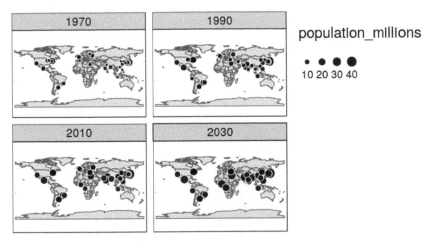

FIGURE 8.13: Faceted map showing the top 30 largest urban agglomerations from 1970 to 2030 based on population projects by the United Nations.

In addition to their utility for showing changing spatial relationships, faceted maps are also useful as the foundation for animated maps (see Section 8.3).

8.2.7 Inset maps

An inset map is a smaller map rendered within or next to the main map. It could serve many different purposes, including providing a context (Figure 8.14) or bringing some non-contiguous regions closer to ease their comparison (Figure 8.15). They could be also used to focus on a smaller area in more detail or to cover the same area as the map, but representing a different topic.

In the example below, we create a map of the central part of New Zealand's Southern Alps. Our inset map will show where the main map is in relation to the whole New Zealand. The first step is to define the area of interest, which can be done by creating a new spatial object, `nz_region`.

```
nz_region = st_bbox(c(xmin = 1340000, xmax = 1450000,
                      ymin = 5130000, ymax = 5210000),
                    crs = st_crs(nz_height)) %>%
  st_as_sfc()
```

In the second step, we create a base map showing the New Zealand's Southern Alps area. This is a place where the most important message is stated.

```
nz_height_map = tm_shape(nz_elev, bbox = nz_region) +
  tm_raster(style = "cont", palette = "YlGn", legend.show = TRUE) +
  tm_shape(nz_height) + tm_symbols(shape = 2, col = "red", size = 1) +
  tm_scale_bar(position = c("left", "bottom"))
```

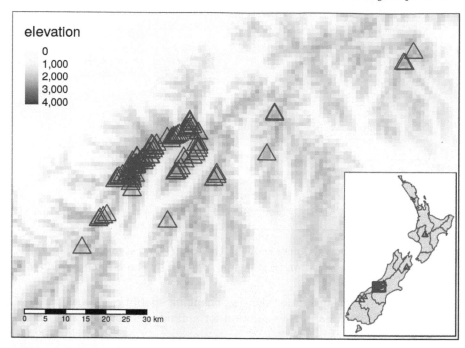

FIGURE 8.14: Inset map providing a context - location of the central part of the Southern Alps in New Zealand.

The third step consists of the inset map creation. It gives a context and helps to locate the area of interest. Importantly, this map needs to clearly indicate the location of the main map, for example by stating its borders.

```
nz_map = tm_shape(nz) + tm_polygons() +
  tm_shape(nz_height) + tm_symbols(shape = 2, col = "red", size = 0.1) +
  tm_shape(nz_region) + tm_borders(lwd = 3)
```

Finally, we combine the two maps using the function viewport() from the **grid** package, the first arguments of which specify the center location (x and y) and a size (width and height) of the inset map.

```
library(grid)
nz_height_map
print(nz_map, vp = viewport(0.8, 0.27, width = 0.5, height = 0.5))
```

Inset map can be saved to file either by using a graphic device (see Section 7.8) or the tmap_save() function and its arguments - insets_tm and insets_vp.

Inset maps are also used to create one map of non-contiguous areas. Probably, the most often used example is a map of the United States, which consists of the contiguous United States, Hawaii and Alaska. It is very important to find the best projection for each individual inset in these types of cases (see Chapter 6 to learn more). We can use US National Atlas Equal Area for the map of the contiguous United States by putting its EPSG code in the `projection` argument of `tm_shape()`.

```
us_states_map = tm_shape(us_states, projection = 2163) + tm_polygons() +
  tm_layout(frame = FALSE)
```

The rest of our objects, `hawaii` and `alaska`, already have proper projections; therefore, we just need to create two separate maps:

```
hawaii_map = tm_shape(hawaii) + tm_polygons() +
  tm_layout(title = "Hawaii", frame = FALSE, bg.color = NA,
            title.position = c("LEFT", "BOTTOM"))
alaska_map = tm_shape(alaska) + tm_polygons() +
  tm_layout(title = "Alaska", frame = FALSE, bg.color = NA)
```

The final map is created by combining and arranging these three maps:

```
us_states_map
print(hawaii_map, vp = grid::viewport(0.35, 0.1, width = 0.2, height = 0.1))
print(alaska_map, vp = grid::viewport(0.15, 0.15, width = 0.3, height = 0.3))
```

The code presented above is compact and can be used as the basis for other inset maps but the results, in Figure 8.15, provide a poor representation of the locations of Hawaii and Alaska. For a more in-depth approach, see the us-map[9] vignette from the **geocompkg.**

8.3   Animated maps

Faceted maps, described in Section 8.2.6, can show how spatial distributions of variables change (e.g., over time), but the approach has disadvantages. Facets become tiny when there are many of them. Furthermore, the fact that each facet is physically separated on the screen or page means that subtle differences between facets can be hard to detect.

Animated maps solve these issues. Although they depend on digital publication, this is becoming less of an issue as more and more content moves online.

---

[9]https://geocompr.github.io/geocompkg/articles/us-map.html

FIGURE 8.15: Map of the United States.

Animated maps can still enhance paper reports: you can always link readers to a web-page containing an animated (or interactive) version of a printed map to help make it come alive. There are several ways to generate animations in R, including with animation packages such as **gganimate**, which builds on **ggplot2** (see Section 8.6). This section focusses on creating animated maps with **tmap** because its syntax will be familiar from previous sections and the flexibility of the approach.

Figure 8.16 is a simple example of an animated map. Unlike the faceted plot, it does not squeeze multiple maps into a single screen and allows the reader to see how the spatial distribution of the world's most populous agglomerations evolve over time (see the book's website for the animated version).

The animated map illustrated in Figure 8.16 can be created using the same **tmap** techniques that generate faceted maps, demonstrated in Section 8.2.6. There are two differences, however, related to arguments in `tm_facets()`:

- `along = "year"` is used instead of `by = "year"`.
- `free.coords = FALSE`, which maintains the map extent for each map iteration.

These additional arguments are demonstrated in the subsequent code chunk:

```
urb_anim = tm_shape(world) + tm_polygons() +
  tm_shape(urban_agglomerations) + tm_dots(size = "population_millions") +
  tm_facets(along = "year", free.coords = FALSE)
```

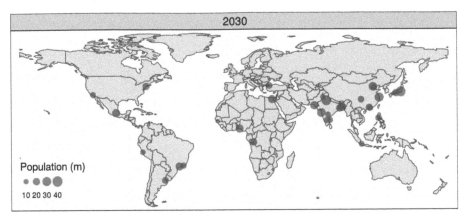

FIGURE 8.16: Animated map showing the top 30 largest urban agglomerations from 1950 to 2030 based on population projects by the United Nations. Animated version available online at: geocompr.robinlovelace.net.

The resulting `urb_anim` represents a set of separate maps for each year. The final stage is to combine them and save the result as a `.gif` file with `tmap_animation()`. The following command creates the animation illustrated in Figure 8.16, with a few elements missing, that we will add in during the exercises:

```
tmap_animation(urb_anim, filename = "urb_anim.gif", delay = 25)
```

Another illustration of the power of animated maps is provided in Figure 8.17. This shows the development of states in the United States, which first formed in the east and then incrementally to the west and finally into the interior. Code to reproduce this map can be found in the script `08-usboundaries.R`.

8.4 Interactive maps

While static and animated maps can enliven geographic datasets, interactive maps can take them to a new level. Interactivity can take many forms, the most common and useful of which is the ability to pan around and zoom into any part of a geographic dataset overlaid on a 'web map' to show context. Less advanced interactivity levels include popups which appear when you click on different features, a kind of interactive label. More advanced levels of interactivity include the ability to tilt and rotate maps, as demonstrated in the **mapdeck** example below, and the provision of "dynamically linked" sub-plots which automatically update when the user pans and zooms (Pezanowski et al., 2018).

FIGURE 8.17: Animated map showing population growth and state formation and boundary changes in the United States, 1790-2010. Animated version available online at geocompr.robinlovelace.net.

The most important type of interactivity, however, is the display of geographic data on interactive or 'slippy' web maps. The release of the **leaflet** package in 2015 revolutionized interactive web map creation from within R and a number of packages have built on these foundations adding new features (e.g., **leaflet.extras**) and making the creation of web maps as simple as creating static maps (e.g., **mapview** and **tmap**). This section illustrates each approach in the opposite order. We will explore how to make slippy maps with **tmap** (the syntax of which we have already learned), **mapview** and finally **leaflet** (which provides low-level control over interactive maps).

A unique feature of **tmap** mentioned in Section 8.2 is its ability to create static and interactive maps using the same code. Maps can be viewed interactively at any point by switching to view mode, using the command `tmap_mode("view")`. This is demonstrated in the code below, which creates an interactive map of New Zealand based on the `tmap` object `map_nz`, created in Section 8.2.2, and illustrated in Figure 8.18:

```
tmap_mode("view")
map_nz
```

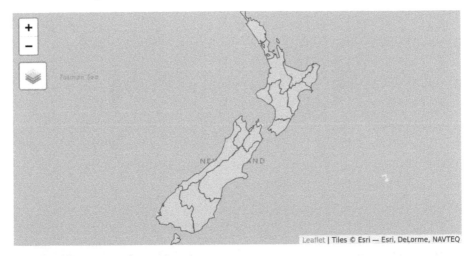

FIGURE 8.18: Interactive map of New Zealand created with tmap in view mode. Interactive version available online at geocompr.robinlovelace.net.

Now that the interactive mode has been 'turned on', all maps produced with **tmap** will launch (another way to create interactive maps is with the `tmap_leaflet` function). Notable features of this interactive mode include the ability to specify the basemap with `tm_basemap()` (or `tmap_options()`) as demonstrated below (result not shown):

```
map_nz + tm_basemap(server = "OpenTopoMap")
```

An impressive and little-known feature of **tmap**'s view mode is that it also works with faceted plots. The argument `sync` in `tm_facets()` can be used in this case to produce multiple maps with synchronized zoom and pan settings, as illustrated in Figure 8.19, which was produced by the following code:

```
world_coffee = left_join(world, coffee_data, by = "name_long")
facets = c("coffee_production_2016", "coffee_production_2017")
tm_shape(world_coffee) + tm_polygons(facets) +
  tm_facets(nrow = 1, sync = TRUE)
```

Switch **tmap** back to plotting mode with the same function:

```
tmap_mode("plot")
#> tmap mode set to plotting
```

If you are not proficient with **tmap**, the quickest way to create interactive

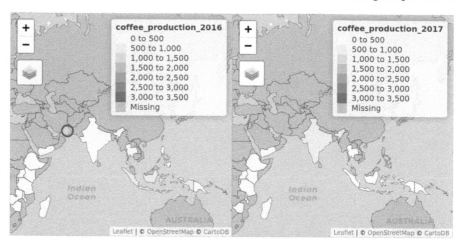

FIGURE 8.19: Faceted interactive maps of global coffee production in 2016 and 2017 in sync, demonstrating tmap's view mode in action.

maps may be with **mapview**. The following 'one liner' is a reliable way to interactively explore a wide range of geographic data formats:

```
mapview::mapview(nz)
```

mapview has a concise syntax yet is powerful. By default, it provides some standard GIS functionality such as mouse position information, attribute queries (via pop-ups), scale bar, and zoom-to-layer buttons. It offers advanced controls including the ability to 'burst' datasets into multiple layers and the addition of multiple layers with + followed by the name of a geographic object. Additionally, it provides automatic coloring of attributes (via argument zcol). In essence, it can be considered a data-driven **leaflet** API (see below for more information about **leaflet**). Given that **mapview** always expects a spatial object (sf, Spatial*, Raster*) as its first argument, it works well at the end of piped expressions. Consider the following example where **sf** is used to intersect lines and polygons and then is visualised with **mapview** (Figure 8.20).

```
trails %>%
  st_transform(st_crs(franconia)) %>%
  st_intersection(franconia[franconia$district == "Oberfranken", ]) %>%
  st_collection_extract("LINE") %>%
  mapview(color = "red", lwd = 3, layer.name = "trails") +
  mapview(franconia, zcol = "district", burst = TRUE) +
  breweries
```

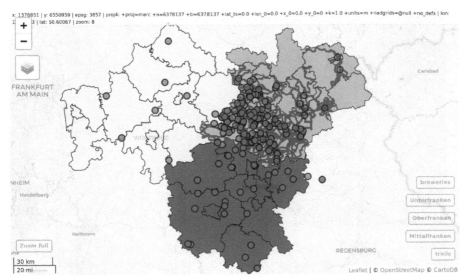

FIGURE 8.20: Using mapview at the end of a sf based pipe expression.

One important thing to keep in mind is that **mapview** layers are added via the + operator (similar to **ggplot2** or **tmap**). This is a frequent gotcha[10] in piped workflows where the main binding operator is `%>%`. For further information on **mapview**, see the package's website at: r-spatial.github.io/mapview/[11].

There are other ways to create interactive maps with R. The **googleway** package, for example, provides an interactive mapping interface that is flexible and extensible (see the `googleway-vignette`[12] for details). Another approach by the same author is **mapdeck**[13], which provides access to Uber's `Deck.gl` framework. Its use of WebGL enables it to interactively visualize large datasets (up to millions of points). The package uses Mapbox access tokens[14], which you must register for before using the package.

Note that the following block assumes the access token is stored in your R environment as `MAPBOX=your_unique_key`. This can be added with `edit_r_environ()` from the **usethis** package.

A unique feature of **mapdeck** is its provision of interactive '2.5d' perspectives, illustrated in Figure 8.21. This means you can can pan, zoom and rotate around the maps, and view the data 'extruded' from the map. Figure 8.21,

---

[10]https://en.wikipedia.org/wiki/Gotcha_(programming)

[11]https://r-spatial.github.io/mapview/articles/

[12]https://cran.r-project.org/web/packages/googleway/vignettes/googleway-vignette.html

[13]https://github.com/SymbolixAU/mapdeck

[14]https://www.mapbox.com/help/how-access-tokens-work/

FIGURE 8.21: Map generated by mapdeck, representing road traffic casualties across the UK. Height of 1 km cells represents number of crashes.

generated by the following code chunk, visualizes road traffic crashes in the UK, with bar height respresenting casualties per area.

```
library(mapdeck)
set_token(Sys.getenv("MAPBOX"))
df = read.csv("https://git.io/geocompr-mapdeck")
ms = mapdeck_style("dark")
mapdeck(style = ms, pitch = 45, location = c(0, 52), zoom = 4) %>%
add_grid(data = df, lat = "lat", lon = "lng", cell_size = 1000,
          elevation_scale = 50, layer_id = "grid_layer",
          colour_range = viridisLite::plasma(5))
```

In the browser you can zoom and drag, in addition to rotating and tilting the map when pressing cmd/ctrl. Multiple layers can be added with the %>% operator, as demonstrated in the mapdeck vignette[15].

Mapdeck also supports sf objects, as can be seen by replacing the add_grid() function call in the preceding code chunk with add_polygon(data = lnd, layer_id = "polygon_layer"), to add polygons representing London to an interactive tilted map.

Last but not least is **leaflet** which is the most mature and widely used interactive mapping package in R. **leaflet** provides a relatively low-level interface to the Leaflet JavaScript library and many of its arguments can be under-

---

[15]https://cran.r-project.org/web/packages/mapdeck/vignettes/mapdeck.html

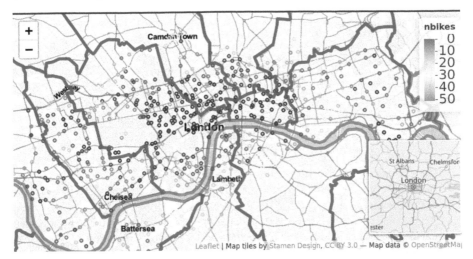

FIGURE 8.22: The leaflet package in action, showing cycle hire points in London.

stood by reading the documentation of the original JavaScript library (see leafletjs.com[16]).

Leaflet maps are created with `leaflet()`, the result of which is a `leaflet` map object which can be piped to other **leaflet** functions. This allows multiple map layers and control settings to be added interactively, as demonstrated in the code below which generates Figure 8.22 (see rstudio.github.io/leaflet/[17] for details).

```
pal = colorNumeric("RdYlBu", domain = cycle_hire$nbikes)
leaflet(data = cycle_hire) %>%
  addProviderTiles(providers$Stamen.TonerLite) %>%
  addCircles(col = ~pal(nbikes), opacity = 0.9) %>%
  addPolygons(data = lnd, fill = FALSE) %>%
  addLegend(pal = pal, values = ~nbikes) %>%
  setView(lng = -0.1, 51.5, zoom = 12) %>%
  addMiniMap()
```

---

[16]http://leafletjs.com/reference-1.3.0.html
[17]https://rstudio.github.io/leaflet/

8.5   Mapping applications

The interactive web maps demonstrated in Section 8.4 can go far. Careful selection of layers to display, base-maps and pop-ups can be used to communicate the main results of many projects involving geocomputation. But the web mapping approach to interactivity has limitations:

- Although the map is interactive in terms of panning, zooming and clicking, the code is static, meaning the user interface is fixed.
- All map content is generally static in a web map, meaning that web maps cannot scale to handle large datasets easily.
- Additional layers of interactivity, such a graphs showing relationships between variables and 'dashboards' are difficult to create using the web-mapping approach.

Overcoming these limitations involves going beyond static web mapping and towards geospatial frameworks and map servers. Products in this field include GeoDjango[18] (which extends the Django web framework and is written in Python[19]), MapGuide[20] (a framework for developing web applications, largely written in C++[21]) and GeoServer[22] (a mature and powerful map server written in Java[23]). Each of these (particularly GeoServer) is scalable, enabling maps to be served to thousands of people daily — assuming there is sufficient public interest in your maps! The bad news is that such server-side solutions require much skilled developer time to set-up and maintain, often involving teams of people with roles such as a dedicated geospatial database administrator (DBA[24]).

The good news is that web mapping applications can now be rapidly created using **shiny**, a package for converting R code into interactive web applications. This is thanks to its support for interactive maps via functions such as `render-Leaflet()`, documented on the Shiny integration[25] section of RStudio's **leaflet** website. This section gives some context, teaches the basics of **shiny** from a web mapping perspective and culminates in a full-screen mapping application in less than 100 lines of code.

The way **shiny** works is well documented at shiny.rstudio.com[26]. The two key elements of a **shiny** app reflect the duality common to most web application

---

[18] https://docs.djangoproject.com/en/2.0/ref/contrib/gis/
[19] https://github.com/django/django
[20] https://www.osgeo.org/projects/mapguide-open-source/
[21] https://trac.osgeo.org/mapguide/wiki/MapGuideArchitecture
[22] http://geoserver.org/
[23] https://github.com/geoserver/geoserver
[24] http://wiki.gis.com/wiki/index.php/Database_administrator
[25] https://rstudio.github.io/leaflet/shiny.html
[26] https://shiny.rstudio.com/

development: 'front end' (the bit the user sees) and 'back end' code. In **shiny** apps, these elements are typically created in objects named ui and server within an R script named app.R, which lives in an 'app folder'. This allows web mapping applications to be represented in a single file, such as the coffeeApp/app.R[27] file in the book's GitHub repo.

In **shiny** apps these are often split into ui.R (short for user interface) and server.R files, naming conventions used by shiny-server, a server-side Linux application for serving shiny apps on public-facing websites. shiny-server also serves apps defined by a single app.R file in an 'app folder'. Learn more at: https://github.com/rstudio/shiny-server.

Before considering large apps, it is worth seeing a minimal example, named 'lifeApp', in action.[28] The code below defines and launches — with the command shinyApp() — a lifeApp, which provides an interactive slider allowing users to make countries appear with progressively lower levels of life expectancy (see Figure 8.23):

```
library(shiny)    # for shiny apps
library(leaflet)  # renderLeaflet function
library(spData)   # loads the world dataset
ui = fluidPage(
  sliderInput(inputId = "life", "Life expectancy", 49, 84, value = 80),
    leafletOutput(outputId = "map")
  )
server = function(input, output) {
  output$map = renderLeaflet({
    leaflet() %>% addProviderTiles("OpenStreetMap.BlackAndWhite") %>%
      addPolygons(data = world[world$lifeExp < input$life, ])})
}
shinyApp(ui, server)
```

The **user interface** (ui) of lifeApp is created by fluidPage(). This contains input and output 'widgets' — in this case, a sliderInput() (many other *Input() functions are available) and a leafletOutput(). These are arranged row-wise by default, explaining why the slider interface is placed directly above the map in Figure 8.23 (see ?column for adding content column-wise).

The **server side** (server) is a function with input and output arguments. output is a list of objects containing elements generated by render*() function — renderLeaflet() which in this example generates output$map. Input elements such

---

[27]https://github.com/Robinlovelace/geocompr/blob/master/coffeeApp/app.R
[28]The word 'app' in this context refers to 'web application' and should not be confused with smartphone apps, the more common meaning of the word.

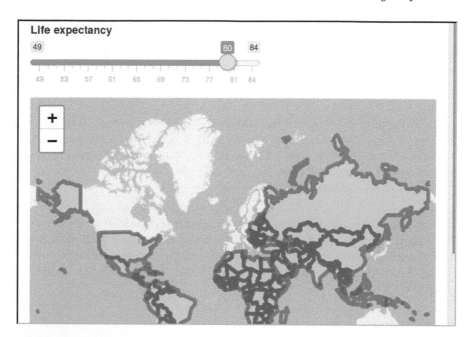

FIGURE 8.23: Screenshot showing minimal example of a web mapping application created with shiny.

as `input$life` referred to in the server must relate to elements that exist in the `ui` — defined by `inputId = "life"` in the code above. The function `shinyApp()` combines both the `ui` and `server` elements and serves the results interactively via a new R process. When you move the slider in the map shown in Figure 8.23, you are actually causing R code to re-run, although this is hidden from view in the user interface.

Building on this basic example and knowing where to find help (see `?shiny`), the best way forward now may be to stop reading and start programming! The recommended next step is to open the previously mentioned `coffeeApp/app.R`[29] script in an IDE of choice, modify it and re-run it repeatedly. The example contains some of the components of a web mapping application implemented in **shiny** and should 'shine' a light on how they behave.

The `coffeeApp/app.R` script contains **shiny** functions that go beyond those demonstrated in the simple 'lifeApp' example. These include `reactive()` and `observe()` (for creating outputs that respond to the user interface — see `?reactive`) and `leafletProxy()` (for modifying a `leaflet` object that has already been created). Such elements are critical to the creation of web mapping applications implemented in **shiny**. A range of 'events' can be programmed

---

[29] https://github.com/Robinlovelace/geocompr/blob/master/coffeeApp/app.R

including advanced functionality such as drawing new layers or subsetting data, as described in the shiny section of RStudio's **leaflet** website.[30]

There are a number of ways to run a **shiny** app. For RStudio users, the simplest way is probably to click on the 'Run App' button located in the top right of the source pane when an `app.R`, `ui.R` or `server.R` script is open. **shiny** apps can also be initiated by using `runApp()` with the first argument being the folder containing the app code and data: `runApp("coffeeApp")` in this case (which assumes a folder named `coffeeApp` containing the `app.R` script is in your working directory). You can also launch apps from a Unix command line with the command `Rscript -e 'shiny::runApp("coffeeApp")'`.

Experimenting with apps such as `coffeeApp` will build not only your knowledge of web mapping applications in R, but also your practical skills. Changing the contents of `setView()`, for example, will change the starting bounding box that the user sees when the app is initiated. Such experimentation should not be done at random, but with reference to relevant documentation, starting with `?shiny`, and motivated by a desire to solve problems such as those posed in the exercises.

shiny used in this way can make prototyping mapping applications faster and more accessible than ever before (deploying **shiny** apps is a separate topic beyond the scope of this chapter). Even if your applications are eventually deployed using different technologies, **shiny** undoubtedly allows web mapping applications to be developed in relatively few lines of code (60 in the case of coffeeApp). That does not stop shiny apps getting rather large. The Propensity to Cycle Tool (PCT) hosted at pct.bike[31], for example, is a national mapping tool funded by the UK's Department for Transport. The PCT is used by dozens of people each day and has multiple interactive elements based on more than 1000 lines of code[32] (Lovelace et al., 2017).

While such apps undoubtedly take time and effort to develop, **shiny** provides a framework for reproducible prototyping that should aid the development process. One potential problem with the ease of developing prototypes with **shiny** is the temptation to start programming too early, before the purpose of the mapping application has been envisioned in detail. For that reason, despite advocating **shiny**, we recommend starting with the longer established technology of a pen and paper as the first stage for interactive mapping projects. This way your prototype web applications should be limited not by technical considerations, but by your motivations and imagination.

---

[30]https://rstudio.github.io/leaflet/shiny.html

[31]http://www.pct.bike/

[32]https://github.com/npct/pct-shiny/blob/master/regions_www/m/server.R

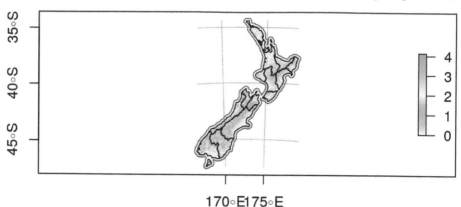

FIGURE 8.24: Map of New Zealand created with plot(). The legend to the right refers to elevation (1000 m above sea level).

8.6   Other mapping packages

tmap provides a powerful interface for creating a wide range of static maps (Section 8.2) and also supports interactive maps (Section 8.4). But there are many other options for creating maps in R. The aim of this section is to provide a taster of some of these and pointers for additional resources: map making is a surprisingly active area of R package development, so there is more to learn than can be covered here.

The most mature option is to use plot() methods provided by core spatial packages **sf** and **raster**, covered in Sections 2.2.3 and 2.3.2, respectively. What we have not mentioned in those sections was that plot methods for raster and vector objects can be combined when the results draw onto the same plot area (elements such as keys in **sf** plots and multi-band rasters will interfere with this). This behavior is illustrated in the subsequent code chunk which generates Figure 8.24. plot() has many other options which can be explored by following links in the ?plot help page and the **sf** vignette sf5[33].

```
g = st_graticule(nz, lon = c(170, 175), lat = c(-45, -40, -35))
plot(nz_water, graticule = g, axes = TRUE, col = "blue")
raster::plot(nz_elev / 1000, add = TRUE)
plot(st_geometry(nz), add = TRUE)
```

Since version 2.3.0[34], the **tidyverse** plotting package **ggplot2** has supported

---

[33]https://cran.r-project.org/web/packages/sf/vignettes/sf5.html
[34]https://www.tidyverse.org/articles/2018/05/ggplot2-2-3-0/

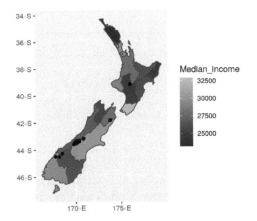

FIGURE 8.25: Map of New Zealand created with ggplot2.

sf objects with `geom_sf()`. The syntax is similar to that used by **tmap:** an initial `ggplot()` call is followed by one or more layers, that are added with `+` `geom_*()`, where `*` represents a layer type such as `geom_sf()` (for sf objects) or `geom_points()` (for points).

ggplot2 plots graticules by default. The default settings for the graticules can be overridden using `scale_x_continuous()`, `scale_y_continuous()` or `coord_sf(datum = NA)`[35]. Other notable features include the use of unquoted variable names encapsulated in `aes()` to indicate which aesthetics vary and switching data sources using the `data` argument, as demonstrated in the code chunk below which creates Figure 8.25:

```
library(ggplot2)
g1 = ggplot() + geom_sf(data = nz, aes(fill = Median_income)) +
  geom_sf(data = nz_height) +
  scale_x_continuous(breaks = c(170, 175))
g1
```

An advantage of **ggplot2** is that it has a strong user-community and many add-on packages. Good additional resources can be found in the open source ggplot2 book[36] (Wickham, 2016) and in the descriptions of the multitude of 'ggpackages' such as **ggrepel** and **tidygraph.**

Another benefit of maps based on **ggplot2** is that they can easily be given a level of interactivity when printed using the function `ggplotly()` from the

---

[35]https://github.com/tidyverse/ggplot2/issues/2071

[36]https://github.com/hadley/ggplot2-book

TABLE 8.1: Selected general-purpose mapping packages.

Package	Title
cartography	Thematic Cartography
ggplot2	Create Elegant Data Visualisations Using the Grammar of Graphics
googleway	Accesses Google Maps APIs to Retrieve Data and Plot Maps
ggspatial	Spatial Data Framework for ggplot2
leaflet	Create Interactive Web Maps with Leaflet
mapview	Interactive Viewing of Spatial Data in R
plotly	Create Interactive Web Graphics via 'plotly.js'
rasterVis	Visualization Methods for Raster Data
tmap	Thematic Maps

plotly package. Try `plotly::ggplotly(g1)`, for example, and compare the result with other **plotly** mapping functions described at: blog.cpsievert.me[37].

At the same time, **ggplot2** has a few drawbacks. The `geom_sf()` function is not always able to create a desired legend to use from the spatial data[38]. Raster objects are also not natively supported in **ggplot2** and need to be converted into a data frame before plotting.

We have covered mapping with **sf**, **raster** and **ggplot2** packages first because these packages are highly flexible, allowing for the creation of a wide range of static maps. Before we cover mapping packages for plotting a specific type of map (in the next paragraph), it is worth considering alternatives to the packages already covered for general-purpose mapping (Table 8.1).

Table 8.1 shows a range of mapping packages are available, and there are many others not listed in this table. Of note is **cartography**, which generates a range of unusual maps including choropleth, 'proportional symbol' and 'flow' maps, each of which is documented in the vignette `cartography`[39].

Several packages focus on specific map types, as illustrated in Table 8.2. Such packages create cartograms that distort geographical space, create line maps, transform polygons into regular or hexagonal grids, and visualize complex data on grids representing geographic topologies.

All of the aforementioned packages, however, have different approaches for data preparation and map creation. In the next paragraph, we focus solely on the

---

[37] https://blog.cpsievert.me/2018/03/30/visualizing-geo-spatial-data-with-sf-and-plotly/

[38] https://github.com/tidyverse/ggplot2/issues/2037

[39] https://cran.r-project.org/web/packages/cartography/vignettes/cartography.html

TABLE 8.2: Selected specific-purpose mapping packages, with associated metrics.

Package	Title
cartogram	Create Cartograms with R
geogrid	Turn Geospatial Polygons into Regular or Hexagonal Grids
geofacet	ggplot2 Faceting Utilities for Geographical Data
globe	Plot 2D and 3D Views of the Earth, Including Major Coastline
linemap	Line Maps

cartogram package. Therefore, we suggest to read the linemap[40], geogrid[41] and geofacet[42] documentations to learn more about them.

A cartogram is a map in which the geometry is proportionately distorted to represent a mapping variable. Creation of this type of map is possible in R with **cartogram**, which allows for creating continuous and non-contiguous area cartograms. It is not a mapping package per se, but it allows for construction of distorted spatial objects that could be plotted using any generic mapping package.

The `cartogram_cont()` function creates continuous area cartograms. It accepts an `sf` object and name of the variable (column) as inputs. Additionally, it is possible to modify the `intermax` argument - maximum number of iterations for the cartogram transformation. For example, we could represent median income in New Zeeland's regions as a continuous cartogram (the right-hand panel of Figure 8.26) as follows:

```
library(cartogram)
nz_carto = cartogram_cont(nz, "Median_income", itermax = 5)
tm_shape(nz_carto) + tm_polygons("Median_income")
```

cartogram also offers creation of non-contiguous area cartograms using `cartogram_ncont()` and Dorling cartograms using `cartogram_dorling()`. Non-contiguous area cartograms are created by scaling down each region based on the provided weighting variable. Dorling cartograms consist of circles with their area proportional to the weighting variable. The code chunk below demonstrates creation of non-contiguous area and Dorling cartograms of US states' population (Figure 8.27):

---

[40] https://github.com/rCarto/linemap
[41] https://github.com/jbaileyh/geogrid
[42] https://github.com/hafen/geofacet

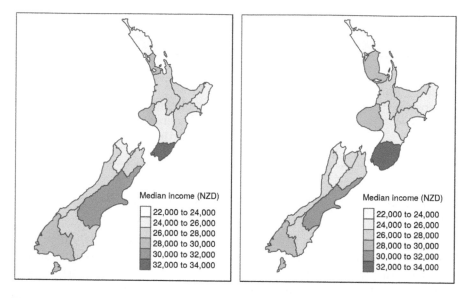

FIGURE 8.26: Comparison of standard map (left) and continuous area cartogram (right).

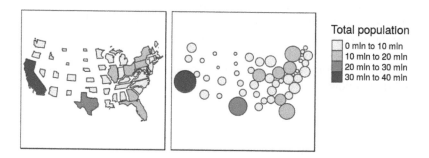

FIGURE 8.27: Comparison of non-continuous area cartogram (left) and Dorling cartogram (right).

```
us_states2163 = st_transform(us_states, 2163)
us_states2163_ncont = cartogram_ncont(us_states2163, "total_pop_15")
us_states2163_dorling = cartogram_dorling(us_states2163, "total_pop_15")
```

New mapping packages are emerging all the time. In 2018 alone, a number of mapping packages have been released on CRAN, including **mapdeck**, **mapsapi**, and **rayshader**. In terms of interactive mapping, **leaflet.extras** contains many functions for extending the functionality of **leaflet** (see the end of the point-pattern[43] vignette in the **geocompkg** website for examples of heatmaps created by **leaflet.extras**).

8.7 Exercises

These exercises rely on a new object, africa. Create it using the world and worldbank_df datasets from the **spData** package as follows (see Chapter 3):

```
africa = world %>%
  filter(continent == "Africa", !is.na(iso_a2)) %>%
  left_join(worldbank_df, by = "iso_a2") %>%
  dplyr::select(name, subregion, gdpPercap, HDI, pop_growth) %>%
  st_transform("+proj=aea +lat_1=20 +lat_2=-23 +lat_0=0 +lon_0=25")
```

We will also use zion and nlcd datasets from **spDataLarge**:

```
zion = st_read((system.file("vector/zion.gpkg", package = "spDataLarge")))
data(nlcd, package = "spDataLarge")
```

1. Create a map showing the geographic distribution of the Human Development Index (HDI) across Africa with base **graphics** (hint: use plot()) and **tmap** packages (hint: use tm_shape(africa) + ...).
   - Name two advantages of each based on the experience.
   - Name three other mapping packages and an advantage of each.
   - Bonus: create three more maps of Africa using these three packages.
2. Extend the **tmap** created for the previous exercise so the legend has three bins: "High" (HDI above 0.7), "Medium" (HDI between 0.55 and 0.7) and "Low" (HDI below 0.55).
   - Bonus: improve the map aesthetics, for example by changing the legend title, class labels and color palette.
3. Represent africa's subregions on the map. Change the default color palette and legend title. Next, combine this map and the map created in the previous exercise into a single plot.
4. Create a land cover map of the Zion National Park.
   - Change the default colors to match your perception of the land cover categories
   - Add a scale bar and north arrow and change the position of both to improve the map's aesthetic appeal

---

[43] https://geocompr.github.io/geocompkg/articles/point-pattern.html

- Bonus: Add an inset map of Zion National Park's location in the context of the Utah state. (Hint: an object representing Utah can be subset from the `us_states` dataset.)

5. Create facet maps of countries in Eastern Africa:
   - With one facet showing HDI and the other representing population growth (hint: using variables `HDI` and `pop_growth`, respectively)
   - With a 'small multiple' per country

6. Building on the previous facet map examples, create animated maps of East Africa:
   - Showing first the spatial distribution of HDI scores then population growth
   - Showing each country in order

7. Create an interactive map of Africa:
   - With **tmap**
   - With **mapview**
   - With **leaflet**
   - Bonus: For each approach, add a legend (if not automatically provided) and a scale bar

8. Sketch on paper ideas for a web mapping app that could be used to make transport or land-use policies more evidence based:
   - In the city you live, for a couple of users per day
   - In the country you live, for dozens of users per day
   - Worldwide for hundreds of users per day and large data serving requirements

9. Update the code in `coffeeApp/app.R` so that instead of centering on Brazil the user can select which country to focus on:
   - Using `textInput()`
   - Using `selectInput()`

10. Reproduce Figure 8.1 and the 1st and 6th panel of Figure 8.6 as closely as possible using the **ggplot2** package.

11. Join `us_states` and `us_states_df` together and calculate a poverty rate for each state using the new dataset. Next, construct a continuous area cartogram based on total population. Finally, create and compare two maps of the poverty rate: (1) a standard choropleth map and (2) a map using the created cartogram boundaries. What is the information provided by the first and the second map? How do they differ from each other?

12. Visualize population growth in Africa. Next, compare it with the maps of a hexagonal and regular grid created using the **geogrid** package.

9

Bridges to GIS software

Prerequisites

- This chapter requires QGIS, SAGA and GRASS to be installed and the following packages to be attached:[1]

```
library(sf)
library(raster)
library(RQGIS)
library(RSAGA)
library(rgrass7)
```

9.1   Introduction

A defining feature of R is the way you interact with it: you type commands and hit Enter (or Ctrl+Enter if writing code in the source editor in RStudio) to execute them interactively. This way of interacting with the computer is called a command-line interface (CLI) (see definition in the note below). CLIs are not unique to R.[2] In dedicated GIS packages, by contrast, the emphasis tends to be on the graphical user interface (GUI). You *can* interact with GRASS, QGIS, SAGA and gvSIG from system terminals and embedded CLIs such as the Python Console in QGIS[3], but 'pointing and clicking' is the norm. This means

---

[1]Packages that have already been used including **spData**, **spDataLarge** and **dplyr** also need to be installed.

[2]Other 'command-lines' include terminals for interacting with the operating system and other interpreted languages such as Python. Many GISs originated as a CLI: it was only after the widespread uptake of computer mice and high-resolution screens in the 1990s that GUIs became common. GRASS, one of the longest-standing GIS programs, for example, relied primarily on command-line interaction before it gained a sophisticated GUI (Landa, 2008).

[3]https://docs.qgis.org/testing/en/docs/pyqgis_developer_cookbook/intro.html

many GIS users miss out on the advantages of the command-line according to Gary Sherman, creator of QGIS (Sherman, 2008):

With the advent of 'modern' GIS software, most people want to point and click their way through life. That's good, but there is a tremendous amount of flexibility and power waiting for you with the command line. Many times you can do something on the command line in a fraction of the time you can do it with a GUI.

The 'CLI vs GUI' debate can be adversial but it does not have to be; both options can be used interchangeably, depending on the task at hand and the user's skillset.[4] The advantages of a good CLI such as that provided by R (and enhanced by IDEs such as RStudio) are numerous. A good CLI

- Facilitates the automation of repetitive tasks.
- Enables transparency and reproducibility, the backbone of good scientific practice and data science.
- Encourages software development by providing tools to modify existing functions and implement new ones.
- Helps develop future-proof programming skills which are in high demand in many disciplines and industries.
- Is user-friendly and fast, allowing an efficient workflow.

On the other hand, GUI-based GIS systems (particularly QGIS) are also advantageous. A good GIS GUI

- Has a 'shallow' learning curve meaning geographic data can be explored and visualized without hours of learning a new language.
- Provides excellent support for 'digitizing' (creating new vector datasets), including trace, snap and topological tools.[5]
- Enables georeferencing (matching raster images to existing maps) with ground control points and orthorectification.
- Supports stereoscopic mapping (e.g., LiDAR and structure from motion).
- Provides access to geodatabase management systems with object-oriented relational data models, topology and fast (spatial) querying.

Another advantage of dedicated GISs is that they provide access to hundreds

---

[4]GRASS GIS and PostGIS are popular in academia and industry and can be seen as products which buck this trend as they are built around the command-line.

[5]The **mapedit** package allows the quick editing of a few spatial features but not professional, large-scale cartographic digitizing.

of 'geoalgorithms' (computational recipes to solve geographic problems — see Chapter 10). Many of these are unavailable from the R command line, except via 'GIS bridges', the topic of (and motivation for) this chapter.[6]

A command-line interface is a means of interacting with computer programs in which the user issues commands via successive lines of text (command lines). bash in Linux and PowerShell in Windows are common examples. CLIs can be augmented with IDEs such as RStudio for R, which provides code auto-completion and other features to improve the user experience.

R originated as an interface language. Its predecessor S provided access to statistical algorithms in other languages (particularly FORTRAN), but from an intuitive read-evaluate-print loop (REPL) (Chambers, 2016). R continues this tradition with interfaces to numerous languages, notably C++, as described in Chapter 1. R was not designed as a GIS. However, its ability to interface with dedicated GISs gives it astonishing geospatial capabilities. R is well known as a statistical programming language, but many people are unaware of its ability to replicate GIS workflows, with the additional benefits of a (relatively) consistent CLI. Furthermore, R outperforms GISs in some areas of geocomputation, including interactive/animated map making (see Chapter 8) and spatial statistical modeling (see Chapter 11). This chapter focuses on 'bridges' to three mature open source GIS products (see Table 9.1): QGIS (via the package **RQGIS**; Section 9.2), SAGA (via **RSAGA**; Section 9.3) and GRASS (via **rgrass7**; Section 9.4). Though not covered here, it is worth being aware of the interface to ArcGIS, a proprietary and very popular GIS software, via **RPyGeo**.^[By the way, it is also possible to use R from within Desktop GIS software packages. The so-called R-ArcGIS bridge (see https://github.com/R-ArcGIS/r-bridge) allows R to be used from within ArcGIS. One can also use R scripts from within QGIS (see https://docs.qgis.org/2.18/en/docs/training_manual/processing/r_intro.html). Finally, it is also possible to use R from the GRASS GIS command line (see https://grasswiki.osgeo.org/wiki/R_statistics/rgrass7). To complement the R-GIS bridges, the chapter ends with a very brief introduction to interfaces to spatial libraries (Section 9.6.1) and spatial databases (Section 9.6.2).

---

[6]An early use of the term 'bridge' referred to the coupling of R with GRASS (Neteler and Mitasova, 2008). Roger Bivand elaborated on this in his talk, "Bridges between GIS and R", delivered at the 2016 GEOSTAT summer school (see slides at: http://spatial.nhh.no/misc/).

TABLE 9.1: Comparison between three open-source GIS. Hybrid refers to the support of vector and raster operations.

GIS	First release	No. functions	Support
GRASS	1984	>500	hybrid
QGIS	2002	>1000	hybrid
SAGA	2004	>600	hybrid

9.2 (R)QGIS

QGIS is one of the most popular open-source GIS (Table 9.1; Graser and Olaya, 2015). Its main advantage lies in the fact that it provides a unified interface to several other open-source GIS. This means that you have access to GDAL, GRASS and SAGA through QGIS (Graser and Olaya, 2015). To run all these geoalgorithms (frequently more than 1000 depending on your set-up) outside of the QGIS GUI, QGIS provides a Python API. **RQGIS** establishes a tunnel to this Python API through the **reticulate** package. Basically, functions `set_env()` and `open_app()` are doing this. Note that it is optional to run `set_env()` and `open_app()` since all functions depending on their output will run them automatically if needed. Before running **RQGIS**, make sure you have installed QGIS and all its (third-party) dependencies such as SAGA and GRASS. To install **RQGIS** a number of dependencies are required, as described in the `install_guide`[7] vignette, which covers installation on Windows, Linux and Mac. At the time of writing (autumn 2018) RQGIS only supports the Long Term Release[8] (2.18), but support for QGIS 3 is in the pipeline (see RQGIS3[9]).

```
library(RQGIS)
set_env(dev = FALSE)
#> $'root'
#> [1] "C:/OSGeo4W64"
#> ...
```

Leaving the `path`-argument of `set_env()` unspecified will search the computer for a QGIS installation. Hence, it is faster to specify explicitly the path to your QGIS installation. Subsequently, `open_app()` sets all paths necessary to run QGIS from within R, and finally creates a so-called QGIS custom application (see http://docs.qgis.org/testing/en/docs/pyqgis_developer_cookbook/intro.html#using-pyqgis-in-custom-applications).

---

[7] https://cran.r-project.org/web/packages/RQGIS/vignettes/install_guide.html
[8] https://qgis.org/en/site/getinvolved/development/roadmap.html
[9] https://github.com/jannes-m/RQGIS3

```
open_app()
```

We are now ready for some QGIS geoprocessing from within R! The example below shows how to unite polygons, a process that unfortunately produces so-called slivers , tiny polygons resulting from overlaps between the inputs that frequently occur in real-world data. We will see how to remove them.

For the union, we use again the incongruent polygons we have already encountered in Section 4.2.5. Both polygon datasets are available in the **spData** package, and for both we would like to use a geographic CRS (see also Chapter 6).

```
data("incongruent", "aggregating_zones", package = "spData")
incongr_wgs = st_transform(incongruent, 4326)
aggzone_wgs = st_transform(aggregating_zones, 4326)
```

To find an algorithm to do this work, `find_algorithms()` searches all QGIS geoalgorithms using regular expressions. Assuming that the short description of the function contains the word "union", we can run:

```
find_algorithms("union", name_only = TRUE)
#> [1] "qgis:union"          "saga:fuzzyunionor" "saga:union"
```

Short descriptions for each geoalgorithm can also be provided, by setting `name_only = FALSE`. If one has no clue at all what the name of a geoalgorithm might be, one can leave the `search_term`-argument empty which will return a list of all available QGIS geoalgorithms. You can also find the algorithms in the QGIS online documentation[10].

The next step is to find out how `qgis:union` can be used. `open_help()` opens the online help of the geoalgorithm in question. `get_usage()` returns all function parameters and default values.

```
alg = "qgis:union"
open_help(alg)
get_usage(alg)
#>ALGORITHM: Union
#>  INPUT <ParameterVector>
#>  INPUT2 <ParameterVector>
#>  OUTPUT <OutputVector>
```

Finally, we can let QGIS do the work. Note that the workhorse function

---

[10]https://docs.qgis.org/2.18/en/docs/user_manual/processing_algs/index.html

`run_qgis()` accepts R named arguments, i.e., you can specify the parameter names as returned by `get_usage()` in `run_qgis()` as you would do in any other regular R function. Note also that `run_qgis()` accepts spatial objects residing in R's global environment as input (here: `aggzone_wgs` and `incongr_wgs`). But of course, you could also specify paths to spatial vector files stored on disk. Setting the `load_output` to `TRUE` automatically loads the QGIS output as an **sf**-object into R.

```
union = run_qgis(alg, INPUT = incongr_wgs, INPUT2 = aggzone_wgs,
                 OUTPUT = file.path(tempdir(), "union.shp"),
                 load_output = TRUE)
#> $'OUTPUT'
#> [1] "C:/Users/geocompr/AppData/Local/Temp/RtmpcJlnUx/union.shp"
```

Note that the QGIS union operation merges the two input layers into one layer by using the intersection and the symmetrical difference of the two input layers (which by the way is also the default when doing a union operation in GRASS and SAGA). This is **not** the same as `st_union(incongr_wgs, aggzone_wgs)` (see Exercises)! The QGIS output contains empty geometries and multipart polygons. Empty geometries might lead to problems in subsequent geoprocessing tasks which is why they will be deleted. `st_dimension()` returns `NA` if a geometry is empty, and can therefore be used as a filter.

```
# remove empty geometries
union = union[!is.na(st_dimension(union)), ]
```

Next we convert multipart polygons into single-part polygons (also known as explode geometries or casting). This is necessary for the deletion of sliver polygons later on.

```
# multipart polygons to single polygons
single = st_cast(union, "POLYGON")
```

One way to identify slivers is to find polygons with comparatively very small areas, here, e.g., 25000 m^2 (see blue colored polygons in the left panel of Figure 9.1).

```
# find polygons which are smaller than 25000 m^2
x = 25000
units(x) = "m^2"
single$area = st_area(single)
sub = dplyr::filter(single, area < x)
```

The next step is to find a function that makes the slivers disappear. Assuming the function or its short description contains the word "sliver", we can run:

```
find_algorithms("sliver", name_only = TRUE)
#> [1] "qgis:eliminatesliverpolygons"
```

This returns only one geoalgorithm whose parameters can be accessed with the help of `get_usage()` again.

```
alg = "qgis:eliminatesliverpolygons"
get_usage(alg)
#>ALGORITHM: Eliminate sliver polygons
#>   INPUT <ParameterVector>
#>   KEEPSELECTION <ParameterBoolean>
#>   ATTRIBUTE <parameters from INPUT>
#>   COMPARISON <ParameterSelection>
#>   COMPARISONVALUE <ParameterString>
#>   MODE <ParameterSelection>
#>   OUTPUT <OutputVector>
#>   ...
```

Conveniently, the user does not need to specify each single parameter. In case a parameter is left unspecified, `run_qgis()` will automatically use the corresponding default value as an argument if available. To find out about the default values, run `get_args_man()`.

To remove the slivers, we specify that all polygons with an area less or equal to 25,000 m^2 should be joined to the neighboring polygon with the largest area (see right panel of Figure 9.1).

```
clean = run_qgis("qgis:eliminatesliverpolygons",
                 INPUT = single,
                 ATTRIBUTE = "area",
                 COMPARISON = "<=",
                 COMPARISONVALUE = 25000,
                 OUTPUT = file.path(tempdir(), "clean.shp"),
                 load_output = TRUE)
#> $'OUTPUT'
#> [1] "C:/Users/geocompr/AppData/Local/Temp/RtmpcJlnUx/clean.shp"
```

Other notes

- Leaving the output parameter(s) unspecified saves the resulting QGIS output to a temporary folder created by QGIS. `run_qgis()` prints these paths to the console after successfully running the QGIS engine.

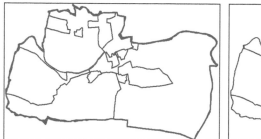 

FIGURE 9.1: Sliver polygons colored in blue (left panel). Cleaned polygons (right panel).

- If the output consists of multiple files and you have set `load_output` to `TRUE`, `run_qgis()` will return a list with each element corresponding to one output file.

To learn more about **RQGIS**, please refer to Muenchow et al. (2017).

9.3 (R)SAGA

The System for Automated Geoscientific Analyses (SAGA; Table 9.1) provides the possibility to execute SAGA modules via the command line interface (`saga_cmd.exe` under Windows and just `saga_cmd` under Linux) (see the SAGA wiki on modules[11]). In addition, there is a Python interface (SAGA Python API). **RSAGA** uses the former to run SAGA from within R.

Though SAGA is a hybrid GIS, its main focus has been on raster processing, and here particularly on digital elevation models (soil properties, terrain attributes, climate parameters). Hence, SAGA is especially good at the fast processing of large (high-resolution) raster datasets (Conrad et al., 2015). Therefore, we will introduce **RSAGA** with a raster use case from Muenchow et al. (2012). Specifically, we would like to compute the SAGA wetness index from a digital elevation model. First of all, we need to make sure that **RSAGA** will find SAGA on the computer when called. For this, all **RSAGA** functions using SAGA in the background make use of `rsaga.env()`. Usually, `rsaga.env()` will detect SAGA automatically by searching several likely directories (see its help for more information).

---

[11] https://sourceforge.net/p/saga-gis/wiki/Executing%20Modules%20with%20SAGA%20CMD/

```
library(RSAGA)
rsaga.env()
```

However, it is possible to have 'hidden' SAGA in a location `rsaga.env()` does not search automatically. `linkSAGA` searches your computer for a valid SAGA installation. If it finds one, it adds the newest version to the PATH environment variable thereby making sure that `rsaga.env()` runs successfully. It is only necessary to run the next code chunk if `rsaga.env()` was unsuccessful (see previous code chunk).

```
library(link2GI)
saga = linkSAGA()
rsaga.env()
```

Secondly, we need to write the digital elevation model to a SAGA-format. Note that calling `data(landslides)` attaches two objects to the global environment - `dem`, a digital elevation model in the form of a `list`, and `landslides`, a `data.frame` containing observations representing the presence or absence of a landslide:

```
data(landslides)
write.sgrd(data = dem, file = file.path(tempdir(), "dem"), header = dem$header)
```

The organization of SAGA is modular. Libraries contain so-called modules, i.e., geoalgorithms. To find out which libraries are available, run:

```
rsaga.get.libraries()
```

We choose the library `ta_hydrology` (ta is the abbreviation for terrain analysis). Subsequently, we can access the available modules of a specific library (here: `ta_hydrology`) as follows:

```
rsaga.get.modules(libs = "ta_hydrology")
```

`rsaga.get.usage()` prints the function parameters of a specific geoalgorithm, e.g., the SAGA Wetness Index, to the console.

```
rsaga.get.usage(lib = "ta_hydrology", module = "SAGA Wetness Index")
```

Finally, you can run SAGA from within R using **RSAGA**'s geoprocessing workhorse function `rsaga.geoprocessor()`. The function expects a parameter-argument list in which you have specified all necessary parameters.

Bridges to GIS software

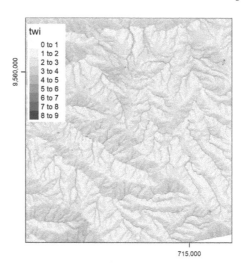

FIGURE 9.2: SAGA wetness index of Mount Mongón, Peru.

```
params = list(DEM = file.path(tempdir(), "dem.sgrd"),
              TWI = file.path(tempdir(), "twi.sdat"))
rsaga.geoprocessor(lib = "ta_hydrology", module = "SAGA Wetness Index",
                   param = params)
```

To facilitate the access to the SAGA interface, **RSAGA** frequently provides user-friendly wrapper-functions with meaningful default values (see **RSAGA** documentation for examples, e.g., ?rsaga.wetness.index). So the function call for calculating the 'SAGA Wetness Index' becomes as simple as:

```
rsaga.wetness.index(in.dem = file.path(tempdir(), "dem"),
                    out.wetness.index = file.path(tempdir(), "twi"))
```

Of course, we would like to inspect our result visually (Figure 9.2). To load and plot the SAGA output file, we use the **raster** package.

```
library(raster)
twi = raster::raster(file.path(tempdir(), "twi.sdat"))
# shown is a version using tmap
plot(twi, col = RColorBrewer::brewer.pal(n = 9, name = "Blues"))
```

You can find an extended version of this example in vignette("RSAGA-landslides") which includes the use of statistical geocomputing to derive terrain attributes as predictors for a non-linear Generalized Additive Model (GAM) to predict spatially landslide susceptibility (Muenchow et al., 2012). The term statistical geocomputation emphasizes the strength of combining R's data science power with the geoprocessing power of a GIS which is at the very heart of building a bridge from R to GIS.

9.4   GRASS through rgrass7

The U.S. Army - Construction Engineering Research Laboratory (USA-CERL) created the core of the Geographical Resources Analysis Support System (GRASS) (Table 9.1; Neteler and Mitasova, 2008) from 1982 to 1995. Academia continued this work since 1997. Similar to SAGA, GRASS focused on raster processing in the beginning while only later, since GRASS 6.0, adding advanced vector functionality (Bivand et al., 2013).

We will introduce **rgrass7** with one of the most interesting problems in GIScience - the traveling salesman problem. Suppose a traveling salesman would like to visit 24 customers. Additionally, he would like to start and finish his journey at home which makes a total of 25 locations while covering the shortest distance possible. There is a single best solution to this problem; however, to find it is even for modern computers (mostly) impossible (Longley, 2015). In our case, the number of possible solutions correspond to `(25 - 1)! / 2`, i.e., the factorial of 24 divided by 2 (since we do not differentiate between forward or backward direction). Even if one iteration can be done in a nanosecond, this still corresponds to 9837145 years. Luckily, there are clever, almost optimal solutions which run in a tiny fraction of this inconceivable amount of time. GRASS GIS provides one of these solutions (for more details, see v.net.salesman[12]). In our use case, we would like to find the shortest path between the first 25 bicycle stations (instead of customers) on London's streets (and we simply assume that the first bike station corresponds to the home of our traveling salesman).

```
data("cycle_hire", package = "spData")
points = cycle_hire[1:25, ]
```

Aside from the cycle hire points data, we will need the OpenStreetMap data of London. We download it with the help of the **osmdata** package (see also Section 7.2). We constrain the download of the street network (in OSM language called "highway") to the bounding box of the cycle hire data, and attach the corresponding data as an `sf`-object. `osmdata_sf()` returns a list with several spatial objects (points, lines, polygons, etc.). Here, we will only keep the line objects. OpenStreetMap objects come with a lot of columns, `streets` features almost 500. In fact, we are only interested in the geometry column.

---

[12]https://grass.osgeo.org/grass77/manuals/v.net.salesman.html

Nevertheless, we are keeping one attribute column; otherwise, we will run into trouble when trying to provide writeVECT() only with a geometry object (see further below and ?writeVECT for more details). Remember that the geometry column is sticky, hence, even though we are just selecting one attribute, the geometry column will be also returned (see Section 2.2.1).

```
library(osmdata)
b_box = st_bbox(points)
london_streets = opq(b_box) %>%
  add_osm_feature(key = "highway") %>%
  osmdata_sf() %>%
  '[['("osm_lines")
london_streets = dplyr::select(london_streets, osm_id)
```

As a convenience to the reader, one can attach london_streets to the global environment using data("london_streets", package = "spDataLarge").

Now that we have the data, we can go on and initiate a GRASS session, i.e., we have to create a GRASS spatial database. The GRASS geodatabase system is based on SQLite. Consequently, different users can easily work on the same project, possibly with different read/write permissions. However, one has to set up this spatial database (also from within R), and users used to a GIS GUI popping up by one click might find this process a bit intimidating in the beginning. First of all, the GRASS database requires its own directory, and contains a location (see the GRASS GIS Database[13] help pages at grass.osgeo.org[14] for further information). The location in turn simply contains the geodata for one project. Within one location, several mapsets can exist and typically refer to different users. PERMANENT is a mandatory mapset and is created automatically. It stores the projection, the spatial extent and the default resolution for raster data. In order to share geographic data with all users of a project, the database owner can add spatial data to the PERMANENT mapset. Please refer to Neteler and Mitasova (2008) and the GRASS GIS quick start[15] for more information on the GRASS spatial database system.

You have to set up a location and a mapset if you want to use GRASS from within R. First of all, we need to find out if and where GRASS 7 is installed on the computer.

```
library(link2GI)
link = findGRASS()
```

---

[13]https://grass.osgeo.org/grass77/manuals/grass_database.html
[14]https://grass.osgeo.org/grass77/manuals/index.html
[15]https://grass.osgeo.org/grass77/manuals/helptext.html

link is a data.frame which contains in its rows the GRASS 7 installations on
your computer. Here, we will use a GRASS 7 installation. If you have not
installed GRASS 7 on your computer, we recommend that you do so now.
Assuming that we have found a working installation on your computer, we
use the corresponding path in initGRASS. Additionally, we specify where to
store the spatial database (gisDbase), name the location london, and use the
PERMANENT mapset.

```
library(rgrass7)
# find a GRASS 7 installation, and use the first one
ind = grep("7", link$version)[1]
# next line of code only necessary if we want to use GRASS as installed by
# OSGeo4W. Among others, this adds some paths to PATH, which are also needed
# for running GRASS.
link2GI::paramGRASSw(link[ind, ])
grass_path =
  ifelse(test = !is.null(link$installation_type) &&
           link$installation_type[ind] == "osgeo4W",
         yes = file.path(link$instDir[ind], "apps/grass", link$version[ind]),
         no = link$instDir)
initGRASS(gisBase = grass_path,
          # home parameter necessary under UNIX-based systems
          home = tempdir(),
          gisDbase = tempdir(), location = "london",
          mapset = "PERMANENT", override = TRUE)
```

Subsequently, we define the projection, the extent and the resolution.

```
execGRASS("g.proj", flags = c("c", "quiet"),
          proj4 = st_crs(london_streets)$proj4string)
b_box = st_bbox(london_streets)
execGRASS("g.region", flags = c("quiet"),
          n = as.character(b_box["ymax"]), s = as.character(b_box["ymin"]),
          e = as.character(b_box["xmax"]), w = as.character(b_box["xmin"]),
          res = "1")
```

Once you are familiar with how to set up the GRASS environment, it becomes
tedious to do so over and over again. Luckily, linkGRASS7() of the **link2GI**
packages lets you do it with one line of code. The only thing you need to
provide is a spatial object which determines the projection and the extent of
the spatial database.. First, linkGRASS7() finds all GRASS installations on your
computer. Since we have set ver_select to TRUE, we can interactively choose
one of the found GRASS-installations. If there is just one installation, the

linkGRASS7() automatically chooses this one. Second, linkGRASS7() establishes a connection to GRASS 7.

```
link2GI::linkGRASS7(london_streets, ver_select = TRUE)
```

Before we can use GRASS geoalgorithms, we need to add data to GRASS's spatial database. Luckily, the convenience function writeVECT() does this for us. (Use writeRast() in the case of raster data.) In our case we add the street and cycle hire point data while using only the first attribute column, and name them also london_streets and points. Note that we are converting the **sf**-objects into objects of class Spatial*. In time, **rgrass7** will also work with **sf**-objects.

```
writeVECT(SDF = as(london_streets, "Spatial"), vname = "london_streets")
writeVECT(SDF = as(points[, 1], "Spatial"), vname = "points")
```

To perform our network analysis, we need a topological clean street network. GRASS's v.clean takes care of the removal of duplicates, small angles and dangles, among others. Here, we break lines at each intersection to ensure that the subsequent routing algorithm can actually turn right or left at an intersection, and save the output in a GRASS object named streets_clean. It is likely that a few of our cycling station points will not lie exactly on a street segment. However, to find the shortest route between them, we need to connect them to the nearest streets segment. v.net's connect-operator does exactly this. We save its output in streets_points_con.

```
# clean street network
execGRASS(cmd = "v.clean", input = "london_streets", output = "streets_clean",
          tool = "break", flags = "overwrite")
# connect points with street network
execGRASS(cmd = "v.net", input = "streets_clean", output = "streets_points_con",
          points = "points", operation = "connect", threshold = 0.001,
          flags = c("overwrite", "c"))
```

The resulting clean dataset serves as input for the v.net.salesman-algorithm, which finally finds the shortest route between all cycle hire stations. center_cats requires a numeric range as input. This range represents the points for which a shortest route should be calculated. Since we would like to calculate the route for all cycle stations, we set it to 1-25. To access the GRASS help page of the traveling salesman algorithm, run execGRASS("g.manual", entry = "v.net.salesman").

```
execGRASS(cmd = "v.net.salesman", input = "streets_points_con",
          output = "shortest_route", center_cats = paste0("1-", nrow(points)),
          flags = c("overwrite"))
```

FIGURE 9.3: Shortest route (blue line) between 24 cycle hire stations (blue dots) on the OSM street network of London.

To visualize our result, we import the output layer into R, convert it into an sf-object keeping only the geometry, and visualize it with the help of the **mapview** package (Figure 9.3 and Section 8.4).

```
route = readVECT("shortest_route") %>%
  st_as_sf() %>%
  st_geometry()
mapview::mapview(route, map.types = "OpenStreetMap.BlackAndWhite", lwd = 7) +
  points
```

Further notes

- Please note that we have used GRASS's spatial database (based on SQLite) which allows faster processing. That means we have only exported geographic data at the beginning. Then we created new objects but only imported the final result back into R. To find out which datasets are currently available, run `execGRASS("g.list", type = "vector,raster", flags = "p")`.
- We could have also accessed an already existing GRASS geodatabase from within R. Prior to importing data into R, you might want to perform some (spatial) subsetting. Use `v.select` and `v.extract` for vector data. `db.select` lets

you select subsets of the attribute table of a vector layer without returning the corresponding geometry.

- You can also start R from within a running GRASS session (for more information please refer to Bivand et al., 2013, and this wiki[16]).
- Refer to the excellent GRASS online help[17] or execGRASS("g.manual", flags = "i") for more information on each available GRASS geoalgorithm.
- If you would like to use GRASS 6 from within R, use the R package **spgrass6**.

9.5   When to use what?

To recommend a single R-GIS interface is hard since the usage depends on personal preferences, the tasks at hand and your familiarity with different GIS software packages which in turn probably depends on your field of study. As mentioned previously, SAGA is especially good at the fast processing of large (high-resolution) raster datasets, and frequently used by hydrologists, climatologists and soil scientists (Conrad et al., 2015). GRASS GIS, on the other hand, is the only GIS presented here supporting a topologically based spatial database which is especially useful for network analyses but also simulation studies (see below). QGIS is much more user-friendly compared to GRASS- and SAGA-GIS, especially for first-time GIS users, and probably the most popular open-source GIS. Therefore, **RQGIS** is an appropriate choice for most use cases. Its main advantages are

- A unified access to several GIS, and therefore the provision of >1000 geoalgorithms (Table 9.1). This includes duplicated functionality, e.g., you can perform overlay-operations using QGIS-, SAGA- or GRASS-geoalgorithms.
- The automatic data format conversions. For instance, SAGA uses .sdat grid files and GRASS uses its own database format but QGIS will handle the corresponding conversions for you on the fly.
- **RQGIS** can also handle spatial objects residing in R as input for geoalgorithms, and loads QGIS output automatically back into R if desired.
- Its convenience functions to support the access of the online help, R named arguments and automatic default value retrieval. Please note that **rgrass7** inspired the latter two features.

By all means, there are use cases when you certainly should use one of the other R-GIS bridges. Though QGIS is the only GIS providing a unified interface to several GIS software packages, it only provides access to a subset of the corresponding third-party geoalgorithms (for more information please refer to Muenchow et al. (2017)). Therefore, to use the complete set of SAGA

---

[16]https://grasswiki.osgeo.org/wiki/R_statistics/rgrass7
[17]https://grass.osgeo.org/grass77/manuals/

and GRASS functions, stick with **RSAGA** and **rgrass7**. When doing so, take advantage of **RSAGA**'s numerous user-friendly functions. Note also, that **RSAGA** offers native R functions for geocomputation such as `multi.local.function()`, `pick.from.points()` and many more. **RSAGA** supports much more SAGA versions than (R)QGIS. Finally, if you need topological correct data and/or spatial database management functionality such as multi-user access, we recommend the usage of GRASS. In addition, if you would like to run simulations with the help of a geodatabase (Krug et al., 2010), use **rgrass7** directly since **RQGIS** always starts a new GRASS session for each call.

Please note that there are a number of further GIS software packages that have a scripting interface but for which there is no dedicated R package that accesses these: gvSig, OpenJump, Orfeo Toolbox and TauDEM.

9.6 Other bridges

The focus of this chapter is on R interfaces to Desktop GIS software. We emphasize these bridges because dedicated GIS software is well-known and a common 'way in' to understanding geographic data. They also provide access to many geoalgorithms.

Other 'bridges' include interfaces to spatial libraries (Section 9.6.1 shows how to access the GDAL CLI from R), spatial databases (see Section 9.6.2) and web mapping services (see Chapter 8). This section provides only a snippet of what is possible. Thanks to R's flexibility, with its ability to call other programs from the system and integration with other languages (notably via **Rcpp** and **reticulate**), many other bridges are possible. The aim is not to be comprehensive, but to demonstrate other ways of accessing the 'flexibility and power' in the quote by Sherman (2008) at the beginning of the chapter.

9.6.1 Bridges to GDAL

As discussed in Chapter 7, GDAL is a low-level library that supports many geographic data formats. GDAL is so effective that most GIS programs use GDAL in the background for importing and exporting geographic data, rather than re-inventing the wheel and using bespoke read-write code. But GDAL offers more than data I/O. It has geoprocessing tools[18] for vector and raster data, functionality to create tiles[19] for serving raster data online, and rapid

---

[18] http://www.gdal.org/pages.html
[19] https://www.gdal.org/gdal2tiles.html

rasterization[20] of vector data, all of which can be accessed via the system of R command line.

The code chunk below demonstrates this functionality: linkGDAL() searches the computer for a working GDAL installation and adds the location of the executable files to the PATH variable, allowing GDAL to be called. In the example below ogrinfo provides metadata on a vector dataset:

```
link2GI::linkGDAL()
cmd = paste("ogrinfo -ro -so -al", system.file("shape/nc.shp", package = "sf"))
system(cmd)
#> INFO: Open of 'C:/Users/geocompr/Documents/R/win-library/3.5/sf/shape/nc.shp'
#>      using driver 'ESRI Shapefile' successful.
#>
#> Layer name: nc
#> Metadata:
#>   DBF_DATE_LAST_UPDATE=2016-10-26
#> Geometry: Polygon
#> Feature Count: 100
#> Extent: (-84.323853, 33.881992) - (-75.456978, 36.589649)
#> Layer SRS WKT:
#> ...
```

This example — which returns the same result as rgdal::ogrInfo() — may be simple, but it shows how to use GDAL via the system command-line, independently of other packages. The 'link' to GDAL provided by **link2gi** could be used as a foundation for doing more advanced GDAL work from the R or system CLI.[21] TauDEM (http://hydrology.usu.edu/taudem/taudem5/index.html) and the Orfeo Toolbox (https://www.orfeo-toolbox.org/) are other spatial data processing libraries/programs offering a command line interface. At the time of writing, it appears that there is only a developer version of an R/TauDEM interface on R-Forge (https://r-forge.r-project.org/R/?group_id=956). In any case, the above example shows how to access these libraries from the system command line via R. This in turn could be the starting point for creating a proper interface to these libraries in the form of new R packages.

Before diving into a project to create a new bridge, however, it is important to be aware of the power of existing R packages and that system() calls may not be platform-independent (they may fail on some computers). Furthermore, **sf** brings most of the power provided by GDAL, GEOS and PROJ to R via the R/C++ interface provided by **Rcpp**, which avoids system() calls.

---

[20] https://www.gdal.org/gdal_rasterize.html
[21] Note also that the **RSAGA** package uses the command line interface to use SAGA geoalgorithms from within R (see Section 9.3).

9.6.2 Bridges to spatial databases

Spatial database management systems (spatial DBMS) store spatial and non-spatial data in a structured way. They can organize large collections of data into related tables (entities) via unique identifiers (primary and foreign keys) and implicitly via space (think for instance of a spatial join). This is useful because geographic datasets tend to become big and messy quite quickly. Databases enable storing and querying large datasets efficiently based on spatial and non-spatial fields, and provide multi-user access and topology support.

The most important open source spatial database is PostGIS (Obe and Hsu, 2015).[22] R bridges to spatial DBMSs such as PostGIS are important, allowing access to huge data stores without loading several gigabytes of geographic data into RAM, and likely crashing the R session. The remainder of this section shows how PostGIS can be called from R, based on "Hello real word" from *PostGIS in Action, Second Edition* (Obe and Hsu, 2015).[23]

The subsequent code requires a working internet connection, since we are accessing a PostgreSQL/PostGIS database which is living in the QGIS Cloud (https://qgiscloud.com/).[24]

```
library(RPostgreSQL)
conn = dbConnect(drv = PostgreSQL(), dbname = "rtafdf_zljbqm",
                 host = "db.qgiscloud.com",
                 port = "5432", user = "rtafdf_zljbqm",
                 password = "d3290ead")
```

Often the first question is, 'which tables can be found in the database?'. This can be asked as follows (the answer is 5 tables):

```
dbListTables(conn)
#> [1] "spatial_ref_sys" "topology"      "layer"         "restaurants"
#> [5] "highways"
```

We are only interested in the restaurants and the highways tables. The former represents the locations of fast-food restaurants in the US, and the latter are principal US highways. To find out about attributes available in a table, we can run:

---

[22]SQLite/SpatiaLite are certainly also important but implicitly we have already introduced this approach since GRASS is using SQLite in the background (see Section 9.4).

[23]Thanks to Manning Publications, Regina Obe and Leo Hsu for permission to use this example.

[24]QGIS Cloud lets you store geographic data and maps in the cloud. In the background, it uses QGIS Server and PostgreSQL/PostGIS. This way, the reader can follow the PostGIS example without the need to have PostgreSQL/PostGIS installed on a local machine. Thanks to the QGIS Cloud team for hosting this example.

```
dbListFields(conn, "highways")
#> [1] "qc_id"           "wkb_geometry" "gid"              "feature"
#> [5] "name"            "state"
```

The first query will select US Route 1 in Maryland (MD). Note that `st_read()` allows us to read geographic data from a database if it is provided with an open connection to a database and a query. Additionally, `st_read()` needs to know which column represents the geometry (here: `wkb_geometry`).

```
query = paste(
  "SELECT *",
  "FROM highways",
  "WHERE name = 'US Route 1' AND state = 'MD';")
us_route = st_read(conn, query = query, geom = "wkb_geometry")
```

This results in an **sf**-object named `us_route` of type `sfc_MULTILINESTRING`. The next step is to add a 20-mile buffer (corresponds to 1609 meters times 20) around the selected highway (Figure 9.4).

```
query = paste(
  "SELECT ST_Union(ST_Buffer(wkb_geometry, 1609 * 20))::geometry",
  "FROM highways",
  "WHERE name = 'US Route 1' AND state = 'MD';")
buf = st_read(conn, query = query)
```

Note that this was a spatial query using functions (`ST_Union()`, `ST_Buffer()`) you should be already familiar with since you find them also in the **sf**-package, though here they are written in lowercase characters (`st_union()`, `st_buffer()`). In fact, function names of the **sf** package largely follow the PostGIS naming conventions.[25] The last query will find all Hardee restaurants (HDE) within the buffer zone (Figure 9.4).

```
query = paste(
  "SELECT r.wkb_geometry",
  "FROM restaurants r",
  "WHERE EXISTS (",
  "SELECT gid",
  "FROM highways",
  "WHERE",
  "ST_DWithin(r.wkb_geometry, wkb_geometry, 1609 * 20) AND",
  "name = 'US Route 1' AND",
```

---

[25] The prefix st stands for space/time.

FIGURE 9.4: Visualization of the output of previous PostGIS commands showing the highway (black line), a buffer (light yellow) and three restaurants (light blue points) within the buffer.

```
"state = 'MD' AND",
"r.franchise = 'HDE');"
)
hardees = st_read(conn, query = query)
```

Please refer to Obe and Hsu (2015) for a detailed explanation of the spatial SQL query. Finally, it is good practice to close the database connection as follows:[26]

```
RPostgreSQL::postgresqlCloseConnection(conn)
```

Unlike PostGIS, **sf** only supports spatial vector data. To query and manipulate raster data stored in a PostGIS database, use the **rpostgis** package (Bucklin and Basille, 2018) and/or use command-line tools such as `rastertopgsql` which comes as part of the PostGIS installation.

This subsection is only a brief introduction to PostgreSQL/PostGIS. Nevertheless, we would like to encourage the practice of storing geographic and non-geographic data in a spatial DBMS while only attaching those subsets to R's global environment which are needed for further (geo-)statistical analysis. Please refer to Obe and Hsu (2015) for a more detailed description of the SQL queries presented and a more comprehensive introduction to PostgreSQL/Post-GIS in general. PostgreSQL/PostGIS is a formidable choice as an open-source spatial database. But the same is true for the lightweight SQLite/SpatiaLite database engine and GRASS which uses SQLite in the background (see Section 9.4).

---

[26]It is important to close the connection here because QGIS Cloud (free version) allows only ten concurrent connections.

As a final note, if your data is getting too big for PostgreSQL/PostGIS and you require massive spatial data management and query performance, then the next logical step is to use large-scale geographic querying on distributed computing systems, as for example, provided by GeoMesa (http://www.geomesa.org/) or GeoSpark (http://geospark.datasyslab.org/; Huang et al., 2017).

9.7   Exercises

1. Create two overlapping polygons (`poly_1` and `poly_2`) with the help of the **sf**-package (see Chapter 2).

2. Union `poly_1` and `poly_2` using `st_union()` and `qgis:union`. What is the difference between the two union operations? How can we use the **sf** package to obtain the same result as QGIS?

3. Calculate the intersection of `poly_1` and `poly_2` using:

   - **RQGIS**, **RSAGA** and **rgrass7**
   - **sf**

4. Attach `data(dem, package = "RQGIS")` and `data(random_points, package = "RQGIS")`. Select randomly a point from `random_points` and find all `dem` pixels that can be seen from this point (hint: viewshed). Visualize your result. For example, plot a hillshade, and on top of it the digital elevation model, your viewshed output and the point. Additionally, give `mapview` a try.

5. Compute catchment area and catchment slope of `data("dem", package = "RQGIS")` using **RSAGA** (see Section 9.3).

6. Use `gdalinfo` via a system call for a raster file stored on disk of your choice (see Section 9.6.1).

7. Query all Californian highways from the PostgreSQL/PostGIS database living in the QGIS Cloud introduced in this chapter (see Section 9.6.2).

10

Scripts, algorithms and functions

Prerequisites

This chapter primarily uses base R; the **sf** package is used to check the result of an algorithm we will develop. It assumes you have an understanding of the geographic classes introduced in Chapter 2 and how they can be used to represent a wide range of input file formats (see Chapter 7).

10.1 Introduction

Chapter 1 established that geocomputation is not only about using existing tools, but developing new ones, "in the form of shareable R scripts and functions". This chapter teaches these building blocks of reproducible code. It also introduces low-level geometric algorithms, of the type used in Chapter 9. Reading it should help you to understand how such algorithms work and to write code that can be used many times, by many people, on multiple datasets. The chapter cannot, by itself, make you a skilled programmer. Programming is hard and requires plenty of practice (Abelson et al., 1996):

> To appreciate programming as an intellectual activity in its own right you must turn to computer programming; you must read and write computer programs — many of them.

There are strong reasons for moving in that direction, however.[1] The advantages of reproducibility go beyond allowing others to replicate your work: reproducible

---

[1]This chapter does not teach programming itself. For more on programming, we recommend Wickham (2014a), Gillespie and Lovelace (2016), and Xiao (2016).

code is often better in every way than code written to be run only once, including in terms of computational efficiency, scalability and ease of adapting and maintaining it.

Scripts are the basis of reproducible R code, a topic covered in Section 10.2. Algorithms are recipes for modifying inputs using a series of steps, resulting in an output, as described in Section 10.3. To ease sharing and reproducibility, algorithms can be placed into functions. That is the topic of Section 10.4. The example of finding the centroid of a polygon will be used to tie these concepts together. Chapter 5 already introduced a centroid function `st_centroid()`, but this example highlights how seemingly simple operations are the result of comparatively complex code, affirming the following observation (Wise, 2001):

One of the most intriguing things about spatial data problems is that things which appear to be trivially easy to a human being can be surprisingly difficult on a computer.

The example also reflects a secondary aim of the chapter which, following Xiao (2016), is "not to duplicate what is available out there, but to show how things out there work".

10.2   Scripts

If functions distributed in packages are the building blocks of R code, scripts are the glue that holds them together, in a logical order, to create reproducible workflows. To programming novices scripts may sound intimidating but they are simply plain text files, typically saved with an extension representing the language they contain. R scripts are generally saved with a `.R` extension and named to reflect what they do. An example is `10-hello.R`, a script file stored in the `code` folder of the book's repository, which contains the following two lines of code:

```
# Aim: provide a minimal R script
print("Hello geocompr")
```

The lines of code may not be particularly exciting but they demonstrate the point: scripts do not need to be complicated. Saved scripts can be called and executed in their entirety with `source()`, as demonstrated below which shows how the comment is ignored but the instruction is executed:

```
source("code/10-hello.R")
#> [1] "Hello geocompr"
```

There are no strict rules on what can and cannot go into script files and nothing to prevent you from saving broken, non-reproducible code.[2] There are, however, some conventions worth following:

- Write the script in order: just like the script of a film, scripts should have a clear order such as 'setup', 'data processing' and 'save results' (roughly equivalent to 'beginning', 'middle' and 'end' in a film).

- Add comments to the script so other people (and your future self) can understand it. At a minimum, a comment should state the purpose of the script (see Figure 10.1) and (for long scripts) divide it into sections. This can be done in RStudio, for example, with the shortcut `Ctrl+Shift+R`, which creates 'foldable' code section headings.

- Above all, scripts should be reproducible: self-contained scripts that will work on any computer are more useful than scripts that only run on your computer, on a good day. This involves attaching required packages at the beginning, reading-in data from persistent sources (such as a reliable website) and ensuring that previous steps have been taken.[3]

It is hard to enforce reproducibility in R scripts, but there are tools that can help. By default, RStudio 'code-checks' R scripts and underlines faulty code with a red wavy line, as illustrated below:

A useful tool for reproducibility is the **reprex** package. Its main function `reprex()` tests lines of R code to check if they are reproducible, and provides markdown output to facilitate communication on sites such as GitHub. See the web page reprex.tidyverse.org for details.

The contents of this section apply to any type of R script. A particular consideration with scripts for geocomputation is that they tend to have external dependencies, such as the QGIS dependency to run code in Chapter 9, and require input data in a specific format. Such dependencies should be mentioned as comments in the script or elsewhere in the project of which it is a part, as illustrated in the script `10-centroid-alg.R`[4]. The work undertaken by this script is demonstrated in the reproducible example below, which works on a

---

[2]Lines of code that do not contain valid R should be commented out, by adding a `#` to the start of the line, to prevent errors. See line 1 of the `10-hello.R` script.

[3]Prior steps can be referred to with a comment or with an if statement such as `if(!exists("x")) source("x.R")` (which would run the script file `x.R` if the object `x` is missing).

[4]https://github.com/Robinlovelace/geocompr/blob/master/code/10-centroid-alg.R

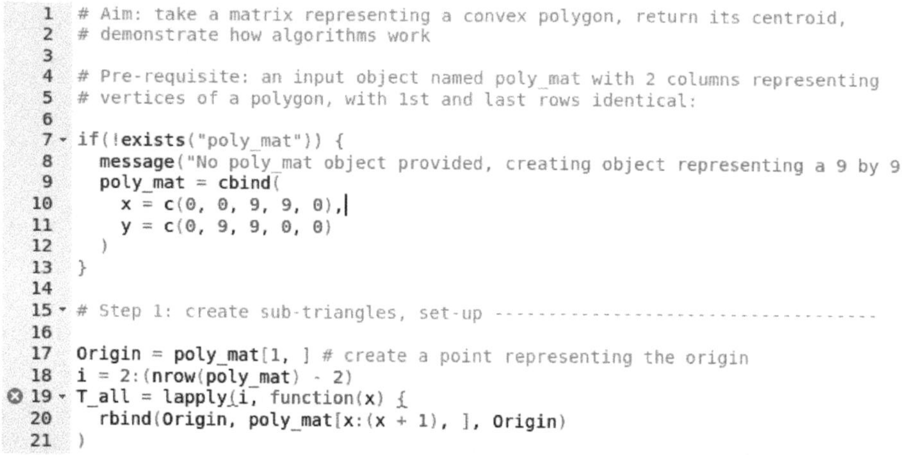

```
 1   # Aim: take a matrix representing a convex polygon, return its centroid,
 2   # demonstrate how algorithms work
 3
 4   # Pre-requisite: an input object named poly_mat with 2 columns representing
 5   # vertices of a polygon, with 1st and last rows identical:
 6
 7 ▾ if(!exists("poly_mat")) {
 8     message("No poly_mat object provided, creating object representing a 9 by 9
 9     poly_mat = cbind(
10       x = c(0, 0, 9, 9, 0),|
11       y = c(0, 9, 9, 0, 0)
12     )
13   }
14
15 ▾ # Step 1: create sub-triangles, set-up --------------------------------------
16
17   Origin = poly_mat[1, ] # create a point representing the origin
18   i = 2:(nrow(poly_mat) - 2)
19 ▾ T_all = lapply(i, function(x) {
20       rbind(Origin, poly_mat[x:(x + 1), ], Origin)
21   )
```

FIGURE 10.1: Code checking in RStudio. This example, from the script 10-centroid-alg.R, highlights an unclosed curly bracket on line 19.

pre-requisite object named `poly_mat`, a square with sides 9 units in length (the meaning of this will become apparent in the next section):[5]

```
poly_mat = cbind(
  x = c(0, 0, 9, 9, 0),
  y = c(0, 9, 9, 0, 0)
)
source("https://git.io/10-centroid-alg.R") # short url
```

```
#> [1] "The area is: 81"
#> [1] "The coordinates of the centroid are: 4.5, 4.5"
```

10.3 Geometric algorithms

Algorithms can be understood as the computing equivalent of a cooking recipe. They are a complete set of instructions which, when undertaken on the input (ingredients), result in useful (tasty) outputs. Before diving into a concrete case study, a brief history will show how algorithms relate to scripts (covered

---

[5]This example shows that `source()` works with URLs (a shortened version is used here), assuming you have an internet connection. If you do not, the same script can be called with `source("code/10-centroid-alg.R")`, assuming you are running R from the root directory of the geocompr folder, which can be downloaded from https://github.com/Robinlovelace/geocompr.

in Section 10.2) and functions (which can be used to generalize algorithms, as we'll see in Section 10.4).

The word "algorithm" originated in 9th century Baghdad with the publication of *Hisab al-jabr w'al-muqabala*, an early math textbook. The book was translated into Latin and became so popular that the author's last name, al-Khwārizmī[6], "was immortalized as a scientific term: Al-Khwarizmi became Alchoarismi, Algorismi and, eventually, algorithm" (Bellos, 2011). In the computing age, algorithm refers to a series of steps that solves a problem, resulting in a pre-defined output. Inputs must be formally defined in a suitable data structure (Wise, 2001). Algorithms often start as flow charts or pseudocode showing the aim of the process before being implemented in code. To ease usability, common algorithms are often packaged inside functions, which may hide some or all of the steps taken (unless you look at the function's source code, see Section 10.4).

Geoalgorithms, such as those we encountered in Chapter 9, are algorithms that take geographic data in and, generally, return geographic results (alternative terms for the same thing include *GIS algorithms* and *geometric algorithms*). That may sound simple but it is a deep subject with an entire academic field, *Computational Geometry*, dedicated to their study (de Berg et al., 2008) and numerous books on the subject. O'Rourke (1998), for example, introduces the subject with a range of progressively harder geometric algorithms using reproducible and freely available C code.

An example of a geometric algorithm is one that finds the centroid of a polygon. There are many approaches to centroid calculation, some of which work only on specific types of spatial data[7]. For the purposes of this section, we choose an approach that is easy to visualize: breaking the polygon into many triangles and finding the centroid of each of these, an approach discussed by Kaiser and Morin (1993) alongside other centroid algorithms (and mentioned briefly in O'Rourke, 1998). It helps to further break down this approach into discrete tasks before writing any code (subsequently referred to as step 1 to step 4, these could also be presented as a schematic diagram or pseudocode):

1. Divide the polygon into contiguous triangles.
2. Find the centroid of each triangle.
3. Find the area of each triangle.
4. Find the area-weighted mean of triangle centroids.

These steps may sound straightforward, but converting words into working code requires some work and plenty of trial-and-error, even when the inputs are constrained: The algorithm will only work for *convex polygons*, which contain no internal angles greater than 180°, no star shapes allowed (packages **decido**

---

[6]https://en.wikipedia.org/wiki/Muhammad_ibn_Musa_al-Khwarizmi

[7]https://en.wikipedia.org/wiki/Centroid

and **sfdct** can triangulate non-convex polygons using external libraries, as shown in the algorithm[8] vignette at geocompr.github.io).

The simplest data structure of a polygon is a matrix of x and y coordinates in which each row represents a vertex tracing the polygon's border in order where the first and last rows are identical (Wise, 2001). In this case, we'll create a polygon with five vertices in base R, building on an example from *GIS Algorithms* (Xiao, 2016, see github.com/gisalgs[9] for Python code), as illustrated in Figure 10.2:

```
# generate a simple matrix representation of a polygon:
x_coords = c(10, 0, 0, 12, 20, 10)
y_coords = c(0, 0, 10, 20, 15, 0)
poly_mat = cbind(x_coords, y_coords)
```

Now that we have an example dataset, we are ready to undertake step 1 outlined above. The code below shows how this can be done by creating a single triangle (T1), that demonstrates the method; it also demonstrates step 2 by calculating its centroid based on the formula[10] $1/3(a + b + c)$ where a to c are coordinates representing the triangle's vertices:

```
# create a point representing the origin:
Origin = poly_mat[1, ]
# create 'triangle matrix':
T1 = rbind(Origin, poly_mat[2:3, ], Origin)
# find centroid (drop = FALSE preserves classes, resulting in a matrix):
C1 = (T1[1, , drop = FALSE] + T1[2, , drop = FALSE] + T1[3, , drop = FALSE]) / 3
```

Step 3 is to find the area of each triangle, so a *weighted mean* accounting for the disproportionate impact of large triangles is accounted for. The formula to calculate the area of a triangle is as follows (Kaiser and Morin, 1993):

$$\frac{Ax(By - Cy) + Bx(Cy - Ay) + Cx(Ay - By)}{2}$$

Where A to C are the triangle's three points and x and y refer to the x and y dimensions. A translation of this formula into R code that works with the data in the matrix representation of a triangle T1 is as follows (the function `abs()` ensures a positive result):

---

[8] https://geocompr.github.io/geocompkg/articles/algorithm.html
[9] https://github.com/gisalgs/geom
[10] https://math.stackexchange.com/q/1702595/

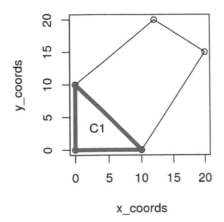

FIGURE 10.2: Illustration of polygon centroid calculation problem.

```
# calculate the area of the triangle represented by matrix T1:
abs(T1[1, 1] * (T1[2, 2] - T1[3, 2]) +
  T1[2, 1] * (T1[3, 2] - T1[1, 2]) +
  T1[3, 1] * (T1[1, 2] - T1[2, 2]) ) / 2
#> [1] 50
```

This code chunk outputs the correct result.[11] The problem is that code is clunky and must by re-typed if we want to run it on another triangle matrix. To make the code more generalizable, we will see how it can be converted into a function in Section 10.4.

Step 4 requires steps 2 and 3 to be undertaken not just on one triangle (as demonstrated above) but on all triangles. This requires *iteration* to create all triangles representing the polygon, illustrated in Figure 10.3. lapply() and vapply() are used to iterate over each triangle here because they provide a concise solution in base R:[12]

```
i = 2:(nrow(poly_mat) - 2)
T_all = lapply(i, function(x) {
  rbind(Origin, poly_mat[x:(x + 1), ], Origin)
})
```

---

[11]The result can be verified with the following formula (which assumes a horizontal base): area is half of the base width times height, $A = B * H/2$. In this case $10 * 10/2 = 50$.

[12]See ?lapply for documentation and Chapter 13 for more on iteration.

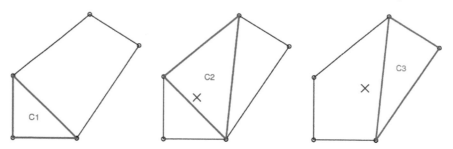

FIGURE 10.3: Illustration of iterative centroid algorithm with triangles. The X represents the area-weighted centroid in iterations 2 and 3.

```
C_list = lapply(T_all,  function(x) (x[1, ] + x[2, ] + x[3, ]) / 3)
C = do.call(rbind, C_list)

A = vapply(T_all, function(x) {
  abs(x[1, 1] * (x[2, 2] - x[3, 2]) +
      x[2, 1] * (x[3, 2] - x[1, 2]) +
      x[3, 1] * (x[1, 2] - x[2, 2]) ) / 2
}, FUN.VALUE = double(1))
```

We are now in a position to complete step 4 to calculate the total area with `sum(A)` and the centroid coordinates of the polygon with `weighted.mean(C[, 1], A)` and `weighted.mean(C[, 2], A)` (exercise for alert readers: verify these commands work). To demonstrate the link between algorithms and scripts, the contents of this section have been condensed into `10-centroid-alg.R`. We saw at the end of Section 10.2 how this script can calculate the centroid of a square. The great thing about *scripting* the algorithm is that it works on the new `poly_mat` object (see exercises below to verify these results with reference to `st_centroid()`):

```
source("code/10-centroid-alg.R")
#> [1] "The area is: 245"
#> [1] "The coordinates of the centroid are: 8.83, 9.22"
```

The example above shows that low-level geographic operations *can* be developed from first principles with base R. It also shows that if a tried-and-tested solution already exists, it may not be worth re-inventing the wheel: if we aimed only to find the centroid of a polygon, it would have been quicker to represent `poly_mat` as an **sf** object and use the pre-existing `sf::st_centroid()` function instead. However, the great benefit of writing algorithms from 1st principles is that you will understand every step of the process, something that cannot be guaranteed when using other peoples' code. A further consideration is performance: R is slow compared with low-level languages such as C++ for number crunching (see Section 1.3) and optimization is difficult. If the aim is to develop new methods, computational efficiency should not be prioritized. This is captured in the saying "premature optimization is the root of all evil (or at least most of it) in programming" (Knuth, 1974).

Algorithm development is hard. This should be apparent from the amount of work that has gone into developing a centroid algorithm in base R that is just one, rather inefficient, approach to the problem with limited real-world applications (convex polygons are uncommon in practice). The experience should lead to an appreciation of low-level geographic libraries such as GEOS (which underlies sf::st_centroid()) and CGAL (the Computational Geometry Algorithms Library) which not only run fast but work on a wide range of input geometry types. A great advantage of the open source nature of such libraries is that their source code is readily available for study, comprehension and (for those with the skills and confidence) modification.[13]

10.4 Functions

Like algorithms, functions take an input and return an output. Functions, however, refer to the implementation in a particular programming language, rather than the 'recipe' itself. In R, functions are objects in their own right, that can be created and joined together in a modular fashion. We can, for example, create a function that undertakes step 2 of our centroid generation algorithm as follows:

```
t_centroid = function(x) {
  (x[1, ] + x[2, ] + x[3, ]) / 3
}
```

The above example demonstrates two key components of functions[14]: 1) the function *body*, the code inside the curly brackets that define what the function does with the inputs; and 2) the *formals*, the list of arguments the function works with — x in this case (the third key component, the environment, is beyond the scope of this section). By default, functions return the last

---

[13]The CGAL function CGAL::centroid() is in fact composed of 7 sub-functions as described at https://doc.cgal.org/latest/Kernel_23/group__centroid__grp.html allowing it to work on a wide range of input data types, whereas the solution we created works only on a very specific input data type. The source code underlying GEOS function Centroid::getCentroid() can be found at https://github.com/libgeos/geos/search?q=getCentroid.

[14]http://adv-r.had.co.nz/Functions.html

object that has been calculated (the coordinates of the centroid in the case of `t_centroid()`).[15]

The function now works on any inputs you pass it, as illustrated in the below command which calculates the area of the 1st triangle from the example polygon in the previous section (see Figure 10.3):

```
t_centroid(T1)
#> x_coords y_coords
#>     3.33     3.33
```

We can also create a function to calculate a triangle's area, which we will name `t_area()`:

```
t_area = function(x) {
  abs(
    x[1, 1] * (x[2, 2] - x[3, 2]) +
    x[2, 1] * (x[3, 2] - x[1, 2]) +
    x[3, 1] * (x[1, 2] - x[2, 2])
  ) / 2
}
```

Note that after the function's creation, a triangle's area can be calculated in a single line of code, avoiding duplication of verbose code: functions are a mechanism for *generalizing* code. The newly created function `t_area()` takes any object x, assumed to have the same dimensions as the 'triangle matrix' data structure we've been using, and returns its area, as illustrated on T1 as follows:

```
t_area(T1)
#> [1] 50
```

We can test the generalizability of the function by using it to find the area of a new triangle matrix, which has a height of 1 and a base of 3:

```
t_new = cbind(x = c(0, 3, 3, 0),
              y = c(0, 0, 1, 0))
t_area(t_new)
#>   x
#> 1.5
```

A useful feature of functions is that they are modular. Provided that you

---

[15]You can also explicitly set the output of a function by adding `return(output)` into the body of the function, where `output` is the result to be returned.

know what the output will be, one function can be used as the building block of another. Thus, the functions `t_centroid()` and `t_area()` can be used as sub-components of a larger function to do the work of the script `10-centroid-alg.R`: calculate the area of any convex polygon. The code chunk below creates the function `poly_centroid()` to mimic the behavior of `sf::st_centroid()` for convex polygons:[16]

```
poly_centroid = function(x) {
  i = 2:(nrow(x) - 2)
  T_all = lapply(i, function(x) {
    rbind(Origin, poly_mat[x:(x + 1), ], Origin)
  })
  C_list = lapply(T_all, t_centroid)
  C = do.call(rbind, C_list)
  A = vapply(T_all, t_area, FUN.VALUE = double(1))
  c(weighted.mean(C[, 1], A), weighted.mean(C[, 2], A))
}
```

```
poly_centroid(poly_mat)
#> [1] 8.83 9.22
```

Functions such as `poly_centroid()` can further be extended to provide different types of output. To return the result as an object of class `sfg`, for example, a 'wrapper' function can be used to modify the output of `poly_centroid()` before returning the result:

```
poly_centroid_sfg = function(x) {
  centroid_coords = poly_centroid(x)
  sf::st_point(centroid_coords)
}
```

We can verify that the output is the same as the output from `sf::st_centroid()` as follows:

```
poly_sfc = sf::st_polygon(list(poly_mat))
identical(poly_centroid_sfg(poly_mat), sf::st_centroid(poly_sfc))
#> [1] TRUE
```

---

[16]Note that the functions we created are called iteratively in `lapply()` and `vapply()` function calls.

10.5 Programming

In this chapter we have moved quickly, from scripts to functions via the tricky topic of algorithms. Not only have we discussed them in the abstract, but we have also created working examples of each to solve a specific problem:

- The script `10-centroid-alg.R` was introduced and demonstrated on a 'polygon matrix'.
- The individual steps that allowed this script to work were described as an algorithm, a computational recipe.
- To generalize the algorithm we converted it into modular functions which were eventually combined to create the function `poly_centroid()` in the previous section.

Taken on its own, each of these steps is straightforward. But the skill of programming is combining scripts, algorithms and functions in a way that produces performant, robust and user-friendly tools that other people can use. If you are new to programming, as we expect most people reading this book will be, being able to follow and reproduce the results in the preceding sections should be seen as a major achievement. Programming takes many hours of dedicated study and practice before you become proficient.

The challenge facing developers aiming to implement new algorithms in an efficient way is put in perspective by considering that we have only created a toy function. In its current state, `poly_centroid()` fails on most (non-convex) polygons! A question arising from this is: how would one generalize the function? Two options are (1) to find ways to triangulate non-convex polygons (a topic covered in the online algorithm[17] article that supports this chapter) and (2) to explore other centroid algorithms that do not rely on triangular meshes.

A wider question is: is it worth programming a solution at all when high performance algorithms have already been implemented and packaged in functions such as `st_centroid()`? The reductionist answer in this specific case is 'no'. In the wider context, and considering the benefits of learning to program, the answer is 'it depends'. With programming, it's easy to waste hours trying to implement a method, only to find that someone has already done the hard work. So instead of seeing this chapter as your first stepping stone towards geometric algorithm programming wizardry, it may be more productive to use it as a lesson in when to try to program a generalized solution, and when to use existing higher-level solutions. There will surely be occasions when writing new functions is the best way forward, but there will also be times when using functions that already exist is the best way forward.

---

[17]`https://geocompr.github.io/geocompkg/articles/algorithm.html`

We cannot guarantee that, having read this chapter, you will be able to rapidly create new functions for your work. But we are confident that its contents will help you decide when is an appropriate time to try (when no other existing functions solve the problem, when the programming task is within your capabilities and when the benefits of the solution are likely to outweigh the time costs of developing it). First steps towards programming can be slow (the exercises below should not be rushed) but the long-term rewards can be large.

10.6 Exercises

1. Read the script `10-centroid-alg.R` in the `code` folder of the book's GitHub repo.
   - Which of the best practices covered in Section 10.2 does it follow?
   - Create a version of the script on your computer in an IDE such as RStudio (preferably by typing-out the script line-by-line, in your own coding style and with your own comments, rather than copy-pasting — this will help you learn how to type scripts). Using the example of a square polygon (e.g., created with `poly_mat = cbind(x = c(0, 0, 9, 9, 0), y = c(0, 9, 9, 0, 0))`) execute the script line-by-line.
   - What changes could be made to the script to make it more reproducible?
   - How could the documentation be improved?
2. In Section 10.3 we calculated that the area and geographic centroid of the polygon represented by `poly_mat` was 245 and 8.8, 9.2, respectively.
   - Reproduce the results on your own computer with reference to the script `10-centroid-alg.R`, an implementation of this algorithm (bonus: type out the commands - try to avoid copy-pasting).
   - Are the results correct? Verify them by converting `poly_mat` into an `sfc` object (named `poly_sfc`) with `st_polygon()` (hint: this function takes objects of class `list()`) and then using `st_area()` and `st_centroid()`.
3. It was stated that the algorithm we created only works for *convex hulls*. Define convex hulls (see Chapter 5) and test the algorithm on a polygon that is *not* a convex hull.
   - Bonus 1: Think about why the method only works for convex hulls and note changes that would need to be made to the algorithm to make it work for other types of polygon.
   - Bonus 2: Building on the contents of `10-centroid-alg.R`, write an

algorithm only using base R functions that can find the total length of linestrings represented in matrix form.

4. In Section 10.4 we created different versions of the `poly_centroid()` function that generated outputs of class `sfg` (`poly_centroid_sfg()`) and type-stable `matrix` outputs (`poly_centroid_type_stable()`). Further extend the function by creating a version (e.g., called `poly_centroid_sf()`) that is type stable (only accepts inputs of class `sf`) *and* returns `sf` objects (hint: you may need to convert the object `x` into a matrix with the command `sf::st_coordinates(x)`).

    • Verify it works by running `poly_centroid_sf(sf::st_sf(sf::st_sfc (poly_sfc)))`

    • What error message do you get when you try to run `poly_centroid_sf(poly_mat)`?

11

Statistical learning

Prerequisites

This chapter assumes proficiency with geographic data analysis, for example gained by studying the contents and working-through the exercises in Chapters 2 to 6. A familiarity with generalized linear models (GLM) and machine learning is highly recommended (for example from Zuur et al., 2009; James et al., 2013).

The chapter uses the following packages:[1]

```
library(sf)
library(raster)
library(mlr)
library(dplyr)
library(parallelMap)
```

Required data will be attached in due course.

11.1 Introduction

Statistical learning is concerned with the use of statistical and computational models for identifying patterns in data and predicting from these patterns. Due to its origins, statistical learning is one of R's great strengths (see Section 1.3).[2] Statistical learning combines methods from statistics and machine learning and its methods can be categorized into supervised and unsupervised techniques.

---

[1] Package **kernlab, pROC, RSAGA** and **spDataLarge** must also be installed although these do not need to be attached.

[2] Applying statistical techniques to geographic data has been an active topic of research for many decades in the fields of Geostatistics, Spatial Statistics and point pattern analysis (Diggle and Ribeiro, 2007; Gelfand et al., 2010; Baddeley et al., 2015).

Both are increasingly used in disciplines ranging from physics, biology and ecology to geography and economics (James et al., 2013).

This chapter focuses on supervised techniques in which there is a training dataset, as opposed to unsupervised techniques such as clustering. Response variables can be binary (such as landslide occurrence), categorical (land use), integer (species richness count) or numeric (soil acidity measured in pH). Supervised techniques model the relationship between such responses — which are known for a sample of observations — and one or more predictors.

The primary aim of much machine learning research is to make good predictions, as opposed to statistical/Bayesian inference, which is good at helping to understand underlying mechanisms and uncertainties in the data (see Krainski et al., 2018). Machine learning thrives in the age of 'big data' because its methods make few assumptions about input variables and can handle huge datasets. Machine learning is conducive to tasks such as the prediction of future customer behavior, recommendation services (music, movies, what to buy next), face recognition, autonomous driving, text classification and predictive maintenance (infrastructure, industry).

This chapter is based on a case study: the (spatial) prediction of landslides. This application links to the applied nature of geocomputation, defined in Chapter 1, and illustrates how machine learning borrows from the field of statistics when the sole aim is prediction. Therefore, this chapter first introduces modeling and cross-validation concepts with the help of a GLM (Zuur et al., 2009). Building on this, the chapter implements a more typical machine learning algorithm, namely a Support Vector Machine (SVM). The models' **predictive performance** will be assessed using spatial cross-validation (CV), which accounts for the fact that geographic data is special.

CV determines a model's ability to generalize to new data, by splitting a dataset (repeatedly) into training and test sets. It uses the training data to fit the model, and checks its performance when predicting against the test data. CV helps to detect overfitting since models that predict the training data too closely (noise) will tend to perform poorly on the test data.

Randomly splitting spatial data can lead to training points that are neighbors in space with test points. Due to spatial autocorrelation, test and training datasets would not be independent in this scenario, with the consequence that CV fails to detect a possible overfitting. Spatial CV alleviates this problem and is the **central** theme in this chapter.

11.2   Case study: Landslide susceptibility

This case study is based on a dataset of landslide locations in Southern Ecuador, illustrated in Figure 11.1 and described in detail in Muenchow et al. (2012). A subset of the dataset used in that paper is provided in the **RSAGA** package, which can be loaded as follows:

```
data("landslides", package = "RSAGA")
```

This should load three objects: a `data.frame` named `landslides`, a `list` named `dem`, and an `sf` object named `study_area`. `landslides` contains a factor column `lslpts` where TRUE corresponds to an observed landslide 'initiation point', with the coordinates stored in columns x and y.[3]

There are 175 landslide points and 1360 non-landslide, as shown by `summary(landslides)`. The 1360 non-landslide points were sampled randomly from the study area, with the restriction that they must fall outside a small buffer around the landslide polygons.

To make the number of landslide and non-landslide points balanced, let us sample 175 from the 1360 non-landslide points.[4]

```
# select non-landslide points
non_pts = filter(landslides, lslpts == FALSE)
# select landslide points
lsl_pts = filter(landslides, lslpts == TRUE)
# randomly select 175 non-landslide points
set.seed(11042018)
non_pts_sub = sample_n(non_pts, size = nrow(lsl_pts))
# create smaller landslide dataset (lsl)
lsl = bind_rows(non_pts_sub, lsl_pts)
```

`dem` is a digital elevation model consisting of two elements: `dem$header`, a `list` which represents a raster 'header' (see Section 2.3), and `dem$data`, a matrix with the altitude of each pixel. `dem` can be converted into a `raster` object with:

---

[3]The landslide initiation point is located in the scarp of a landslide polygon. See Muenchow et al. (2012) for further details.

[4]The `landslides` dataset has been used in classes and summer schools. To show how predictive performance of different algorithms changes with an unbalanced and highly spatially autocorrelated response variable, 1360 non-landslide points were randomly selected, i.e., many more absences than presences. However, especially a logistic regression with a log-link, as used in this chapter, expects roughly the same number of presences and absences in the response.

FIGURE 11.1: Landslide initiation points (red) and points unaffected by landsliding (blue) in Southern Ecuador.

```
dem = raster(
  dem$data,
  crs = dem$header$proj4string,
  xmn = dem$header$xllcorner,
  xmx = dem$header$xllcorner + dem$header$ncols * dem$header$cellsize,
  ymn = dem$header$yllcorner,
  ymx = dem$header$yllcorner + dem$header$nrows * dem$header$cellsize
  )
```

To model landslide susceptibility, we need some predictors. Terrain attributes are frequently associated with landsliding (Muenchow et al., 2012), and these can be computed from the digital elevation model (dem) using R-GIS bridges (see Chapter 9). We leave it as an exercise to the reader to compute the following terrain attribute rasters and extract the corresponding values to our landslide/non-landslide data frame (see exercises; we also provide the resulting data frame via the **spDataLarge** package, see further below):

- slope: slope angle (°).
- cplan: plan curvature (rad m^{-1}) expressing the convergence or divergence of a slope and thus water flow.
- cprof: profile curvature (rad m^{-1}) as a measure of flow acceleration, also known as downslope change in slope angle.

TABLE 11.1: Structure of the lsl dataset.

x	y	lslpts	slope	cplan	cprof	elev	log10_carea
715078	9558647	FALSE	37	0.021	0.009	2500	2.6
713748	9558047	FALSE	42	-0.024	0.007	2500	3.1
712508	9558887	FALSE	20	0.039	0.015	2100	2.3

- `elev`: elevation (m a.s.l.) as the representation of different altitudinal zones of vegetation and precipitation in the study area.
- `log10_carea`: the decadic logarithm of the catchment area (log10 m^2) representing the amount of water flowing towards a location.

Data containing the landslide points, with the corresponding terrain attributes, is provided in the **spDataLarge** package, along with the terrain attribute raster stack from which the values were extracted. Hence, if you have not computed the predictors yourself, attach the corresponding data before running the code of the remaining chapter:

```
# attach landslide points with terrain attributes
data("lsl", package = "spDataLarge")
# attach terrain attribute raster stack
data("ta", package = "spDataLarge")
```

The first three rows of `lsl`, rounded to two significant digits, can be found in Table 11.1.

11.3 Conventional modeling approach in R

Before introducing the **mlr** package, an umbrella-package providing a unified interface to dozens of learning algorithms (Section 11.5), it is worth taking a look at the conventional modeling interface in R. This introduction to supervised statistical learning provides the basis for doing spatial CV, and contributes to a better grasp on the **mlr** approach presented subsequently.

Supervised learning involves predicting a response variable as a function of predictors (Section 11.4). In R, modeling functions are usually specified using formulas (see `?formula` and the detailed Formulas in R Tutorial[5] for details of R formulas). The following command specifies and runs a generalized linear model:

---
[5] https://www.datacamp.com/community/tutorials/r-formula-tutorial

```
fit = glm(lslpts ~ slope + cplan + cprof + elev + log10_carea,
          family = binomial(),
          data = lsl)
```

It is worth understanding each of the three input arguments:

- A formula, which specifies landslide occurrence (lslpts) as a function of the predictors.
- A family, which specifies the type of model, in this case binomial because the response is binary (see ?family).
- The data frame which contains the response and the predictors.

The results of this model can be printed as follows (summary(fit) provides a more detailed account of the results):

```
class(fit)
#> [1] "glm" "lm"
fit
#>
#> Call:  glm(formula = lslpts ~ slope + cplan + cprof + elev + log10_carea,
#>      family = binomial(), data = lsl)
#>
#> Coefficients:
#> (Intercept)       slope        cplan        cprof         elev
#>    1.97e+00    9.30e-02    -2.57e+01    -1.43e+01     2.41e-05
#> log10_carea
#>    -2.12e+00
#>
#> Degrees of Freedom: 349 Total (i.e. Null);  344 Residual
#> Null Deviance:      485
#> Residual Deviance: 361    AIC: 373
```

The model object fit, of class glm, contains the coefficients defining the fitted relationship between response and predictors. It can also be used for prediction. This is done with the generic predict() method, which in this case calls the function predict.glm(). Setting type to response returns the predicted probabilities (of landslide occurrence) for each observation in lsl, as illustrated below (see ?predict.glm):

```
pred_glm = predict(object = fit, type = "response")
head(pred_glm)
#>      1      2      3      4      5      6
#> 0.3327 0.4755 0.0995 0.1480 0.3486 0.6766
```

Spatial predictions can be made by applying the coefficients to the predictor

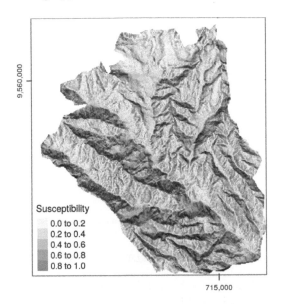

FIGURE 11.2: Spatial prediction of landslide susceptibility using a GLM.

rasters. This can be done manually or with `raster::predict()`. In addition to a model object (`fit`), this function also expects a raster stack with the predictors named as in the model's input data frame (Figure 11.2).

```
# making the prediction
pred = raster::predict(ta, model = fit, type = "response")
```

Here, when making predictions we neglect spatial autocorrelation since we assume that on average the predictive accuracy remains the same with or without spatial autocorrelation structures. However, it is possible to include spatial autocorrelation structures into models (Zuur et al., 2009; Blangiardo and Cameletti, 2015; Zuur et al., 2017) as well as into predictions (kriging approaches, see, e.g., Goovaerts, 1997; Hengl, 2007; Bivand et al., 2013). This is, however, beyond the scope of this book.

Spatial prediction maps are one very important outcome of a model. Even more important is how good the underlying model is at making them since a prediction map is useless if the model's predictive performance is bad. The most popular measure to assess the predictive performance of a binomial model is the Area Under the Receiver Operator Characteristic Curve (AUROC). This is a value between 0.5 and 1.0, with 0.5 indicating a model that is no better than random and 1.0 indicating perfect prediction of the two classes. Thus, the higher the AUROC, the better the model's predictive power. The following code chunk computes the AUROC value of the model with `roc()`, which takes the response and the predicted values as inputs. `auc()` returns the area under the curve.

```
pROC::auc(pROC::roc(lsl$lslpts, fitted(fit)))
#> Area under the curve: 0.826
```

An AUROC value of 0.83 represents a good fit. However, this is an overoptimistic estimation since we have computed it on the complete dataset. To derive a biased-reduced assessment, we have to use cross-validation and in the case of spatial data should make use of spatial CV.

11.4   Introduction to (spatial) cross-validation

Cross-validation belongs to the family of resampling methods (James et al., 2013). The basic idea is to split (repeatedly) a dataset into training and test sets whereby the training data is used to fit a model which then is applied to the test set. Comparing the predicted values with the known response values from the test set (using a performance measure such as the AUROC in the binomial case) gives a bias-reduced assessment of the model's capability to generalize the learned relationship to independent data. For example, a 100-repeated 5-fold cross-validation means to randomly split the data into five partitions (folds) with each fold being used once as a test set (see upper row of Figure 11.3). This guarantees that each observation is used once in one of the test sets, and requires the fitting of five models. Subsequently, this procedure is repeated 100 times. Of course, the data splitting will differ in each repetition. Overall, this sums up to 500 models, whereas the mean performance measure (AUROC) of all models is the model's overall predictive power.

However, geographic data is special. As we will see in Chapter 12, the 'first law' of geography states that points close to each other are, generally, more similar than points further away (Miller, 2004). This means these points are not statistically independent because training and test points in conventional CV are often too close to each other (see first row of Figure 11.3). 'Training' observations near the 'test' observations can provide a kind of 'sneak preview': information that should be unavailable to the training dataset. To alleviate this problem 'spatial partitioning' is used to split the observations into spatially disjointed subsets (using the observations' coordinates in a k-means clustering; Brenning (2012b); second row of Figure 11.3). This partitioning strategy is the **only** difference between spatial and conventional CV. As a result, spatial CV leads to a bias-reduced assessment of a model's predictive performance, and hence helps to avoid overfitting.

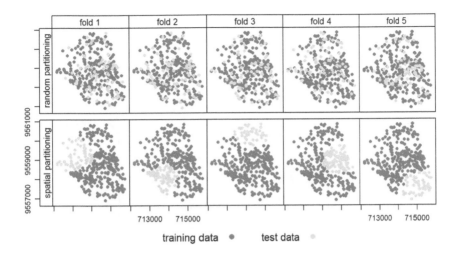

FIGURE 11.3: Spatial visualization of selected test and training observations for cross-validation of one repetition. Random (upper row) and spatial partitioning (lower row).

11.5 Spatial CV with mlr

There are dozens of packages for statistical learning, as described for example in the CRAN machine learning task view[6]. Getting acquainted with each of these packages, including how to undertake cross-validation and hyperparameter tuning, can be a time-consuming process. Comparing model results from different packages can be even more laborious. The **mlr** package was developed to address these issues. It acts as a 'meta-package', providing a unified interface to popular supervised and unsupervised statistical learning techniques including classification, regression, survival analysis and clustering (Bischl et al., 2016). The standardized **mlr** interface is based on eight 'building blocks'. As illustrated in Figure 11.4, these have a clear order.

The **mlr** modeling process consists of three main stages. First, a **task** specifies the data (including response and predictor variables) and the model type (such as regression or classification). Second, a **learner** defines the specific learning algorithm that is applied to the created task. Third, the **resampling** approach assesses the predictive performance of the model, i.e., its ability to generalize to new data (see also Section 11.4).

---

[6]https://CRAN.R-project.org/view=MachineLearning

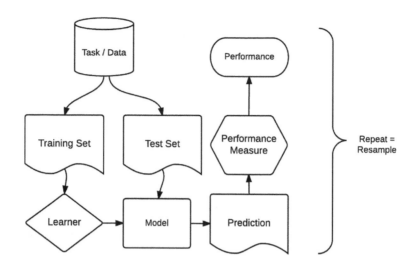

FIGURE 11.4: Basic building blocks of the mlr package. Source: http://bit.ly/2tcb2b7. (Permission to reuse this figure was kindly granted.)

11.5.1 Generalized linear model

To implement a GLM in **mlr**, we must create a **task** containing the landslide data. Since the response is binary (two-category variable), we create a classification task with makeClassifTask() (for regression tasks, use makeRegrTask(), see ?makeRegrTask for other task types). The first essential argument of these make*() functions is data. The target argument expects the name of a response variable and positive determines which of the two factor levels of the response variable indicate the landslide initiation point (in our case this is TRUE). All other variables of the lsl dataset will serve as predictors except for the coordinates (see the result of getTaskFormula(task) for the model formula). For spatial CV, the coordinates parameter is used (see Section 11.4 and Figure 11.3) which expects the coordinates as a xy data frame.

```
library(mlr)
# coordinates needed for the spatial partitioning
coords = lsl[, c("x", "y")]
# select response and predictors to use in the modeling
data = dplyr::select(lsl, -x, -y)
# create task
task = makeClassifTask(data = data, target = "lslpts",
                       positive = "TRUE", coordinates = coords)
```

TABLE 11.2: Sample of available learners for binomial tasks in the mlr package.

Class	Name	Short name	Package
classif.binomial	Binomial Regression	binomial	stats
classif.featureless	Featureless classifier	featureless	mlr
classif.fnn	Fast k-Nearest Neighbour	fnn	FNN
classif.gausspr	Gaussian Processes	gausspr	kernlab
classif.knn	k-Nearest Neighbor	knn	class
classif.ksvm	Support Vector Machines	ksvm	kernlab

makeLearner() determines the statistical learning method to use. All classification **learners** start with classif. and all regression learners with regr. (see ?makeLearners for details). listLearners() helps to find out about all available learners and from which package **mlr** imports them (Table 11.2). For a specific task, we can run:

```
listLearners(task, warn.missing.packages = FALSE) %>%
  dplyr::select(class, name, short.name, package) %>%
  head
```

This yields all learners able to model two-class problems (landslide yes or no). We opt for the binomial classification method used in Section 11.3 and implemented as classif.binomial in **mlr**. Additionally, we must specify the link-function, logit in this case, which is also the default of the binomial() function. predict.type determines the type of the prediction with prob resulting in the predicted probability for landslide occurrence between 0 and 1 (this corresponds to type = response in predict.glm).

```
lrn = makeLearner(cl = "classif.binomial",
                  link = "logit",
                  predict.type = "prob",
                  fix.factors.prediction = TRUE)
```

To find out from which package the specified learner is taken and how to access the corresponding help pages, we can run:

```
getLearnerPackages(lrn)
helpLearner(lrn)
```

The set-up steps for modeling with **mlr** may seem tedious. But remember, this single interface provides access to the 150+ learners shown by listLearners();

it would be far more tedious to learn the interface for each learner! Further advantages are simple parallelization of resampling techniques and the ability to tune machine learning hyperparameters (see Section 11.5.2). Most importantly, (spatial) resampling in **mlr** is straightforward, requiring only two more steps: specifying a resampling method and running it. We will use a 100-repeated 5-fold spatial CV: five partitions will be chosen based on the provided coordinates in our `task` and the partitioning will be repeated 100 times:[7]

```
perf_level = makeResampleDesc(method = "SpRepCV", folds = 5, reps = 100)
```

To execute the spatial resampling, we run `resample()` using the specified learner, task, resampling strategy and of course the performance measure, here the AUROC. This takes some time (around 10 seconds on a modern laptop) because it computes the AUROC for 500 models. Setting a seed ensures the reproducibility of the obtained result and will ensure the same spatial partitioning when re-running the code.

```
set.seed(012348)
sp_cv = mlr::resample(learner = lrn, task = task,
                      resampling = perf_level,
                      measures = mlr::auc)
```

The output of the preceding code chunk is a bias-reduced assessment of the model's predictive performance, as illustrated in the following code chunk (required input data is saved in the file `spatialcv.Rdata` in the book's GitHub repo):

```
# summary statistics of the 500 models
summary(sp_cv$measures.test$auc)
#>    Min. 1st Qu.  Median    Mean 3rd Qu.    Max.
#>   0.686   0.757   0.789   0.780   0.795   0.861
# mean AUROC of the 500 models
mean(sp_cv$measures.test$auc)
#> [1] 0.78
```

To put these results in perspective, let us compare them with AUROC values from a 100-repeated 5-fold non-spatial cross-validation (Figure 11.5; the code for the non-spatial cross-validation is not shown here but will be explored in

---

[7]Note that package **sperrorest** initially implemented spatial cross-validation in R (Brenning, 2012b). In the meantime, its functionality was integrated into the **mlr** package which is the reason why we are using **mlr** (Schratz et al., 2018).The **caret** package is another umbrella-package (Kuhn and Johnson, 2013) for streamlined modeling in R; however, so far it does not provide spatial CV which is why we refrain from using it for spatial data.

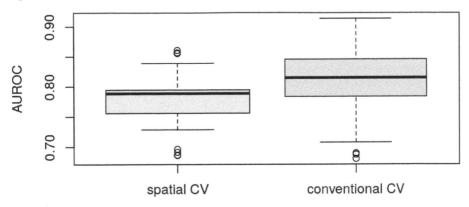

FIGURE 11.5: Boxplot showing the difference in AUROC values between spatial and conventional 100-repeated 5-fold cross-validation.

the exercise section). As expected, the spatially cross-validated result yields lower AUROC values on average than the conventional cross-validation approach, underlining the over-optimistic predictive performance due to spatial autocorrelation of the latter.

11.5.2   Spatial tuning of machine-learning hyperparameters

Section 11.4 introduced machine learning as part of statistical learning. To recap, we adhere to the following definition of machine learning by Jason Brownlee[8]:

Machine learning, more specifically the field of predictive modeling, is primarily concerned with minimizing the error of a model or making the most accurate predictions possible, at the expense of explainability. In applied machine learning we will borrow, reuse and steal algorithms from many different fields, including statistics and use them towards these ends.

In Section 11.5.1 a GLM was used to predict landslide susceptibility. This section introduces support vector machines (SVM) for the same purpose. Random forest models might be more popular than SVMs; however, the positive effect of tuning hyperparameters on model performance is much

---

[8]https://machinelearningmastery.com/linear-regression-for-machine-learning/

more pronounced in the case of SVMs (Probst et al., 2018). Since (spatial) hyperparameter tuning is the major aim of this section, we will use an SVM. For those wishing to apply a random forest model, we recommend to read this chapter, and then proceed to Chapter 14 in which we will apply the currently covered concepts and techniques to make spatial predictions based on a random forest model.

SVMs search for the best possible 'hyperplanes' to separate classes (in a classification case) and estimate 'kernels' with specific hyperparameters to allow for non-linear boundaries between classes (James et al., 2013). Hyperparameters should not be confused with coefficients of parametric models, which are sometimes also referred to as parameters.[9] Coefficients can be estimated from the data, while hyperparameters are set before the learning begins. Optimal hyperparameters are usually determined within a defined range with the help of cross-validation methods. This is called hyperparameter tuning.

Some SVM implementations such as that provided by **kernlab** allow hyperparameters to be tuned automatically, usually based on random sampling (see upper row of Figure 11.3). This works for non-spatial data but is of less use for spatial data where 'spatial tuning' should be undertaken.

Before defining spatial tuning, we will set up the **mlr** building blocks, introduced in Section 11.5.1, for the SVM. The classification task remains the same, hence we can simply reuse the `task` object created in Section 11.5.1. Learners implementing SVM can be found using `listLearners()` as follows:

```
lrns = listLearners(task, warn.missing.packages = FALSE)
filter(lrns, grepl("svm", class)) %>%
  dplyr::select(class, name, short.name, package)
#>            class                                  name short.name package
#> 6    classif.ksvm                Support Vector Machines       ksvm kernlab
#> 9   classif.lssvm Least Squares Support Vector Machine      lssvm kernlab
#> 17    classif.svm     Support Vector Machines (libsvm)        svm   e1071
```

Of the options illustrated above, we will use `ksvm()` from the **kernlab** package (Karatzoglou et al., 2004). To allow for non-linear relationships, we use the popular radial basis function (or Gaussian) kernel which is also the default of `ksvm()`.

```
lrn_ksvm = makeLearner("classif.ksvm",
                        predict.type = "prob",
                        kernel = "rbfdot")
```

---

[9]For a detailed description of the difference between coefficients and hyperparameters, see the 'machine mastery' blog post on the subject.

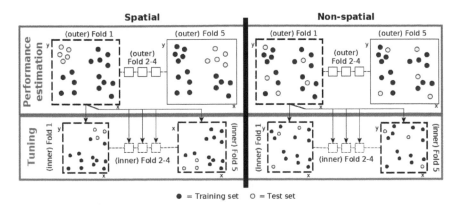

FIGURE 11.6: Schematic of hyperparameter tuning and performance estimation levels in CV. (Figure was taken from Schratz et al. (2018). Permission to reuse it was kindly granted.)

The next stage is to specify a resampling strategy. Again we will use a 100-repeated 5-fold spatial CV.

```
# performance estimation level
perf_level = makeResampleDesc(method = "SpRepCV", folds = 5, reps = 100)
```

Note that this is the exact same code as used for the GLM in Section 11.5.1; we have simply repeated it here as a reminder.

So far, the process has been identical to that described in Section 11.5.1. The next step is new, however: to tune the hyperparameters. Using the same data for the performance assessment and the tuning would potentially lead to overoptimistic results (Cawley and Talbot, 2010). This can be avoided using nested spatial CV.

This means that we split each fold again into five spatially disjoint subfolds which are used to determine the optimal hyperparameters (tune_level object in the code chunk below; see Figure 11.6 for a visual representation). To find the optimal hyperparameter combination, we fit 50 models (ctrl object in the code chunk below) in each of these subfolds with randomly selected values for the hyperparameters C and Sigma. The random selection of values C and Sigma is additionally restricted to a predefined tuning space (ps object). The range of the tuning space was chosen with values recommended in the literature (Schratz et al., 2018).

```
# five spatially disjoint partitions
tune_level = makeResampleDesc("SpCV", iters = 5)
# use 50 randomly selected hyperparameters
ctrl = makeTuneControlRandom(maxit = 50)
# define the outer limits of the randomly selected hyperparameters
ps = makeParamSet(
  makeNumericParam("C", lower = -12, upper = 15, trafo = function(x) 2^x),
  makeNumericParam("sigma", lower = -15, upper = 6, trafo = function(x) 2^x)
  )
```

The next stage is to modify the learner lrn_ksvm in accordance with all the characteristics defining the hyperparameter tuning with makeTuneWrapper().

```
wrapped_lrn_ksvm = makeTuneWrapper(learner = lrn_ksvm,
                                   resampling = tune_level,
                                   par.set = ps,
                                   control = ctrl,
                                   show.info = TRUE,
                                   measures = mlr::auc)
```

The **mlr** is now set-up to fit 250 models to determine optimal hyperparameters for one fold. Repeating this for each fold, we end up with 1250 (250 * 5) models for each repetition. Repeated 100 times means fitting a total of 125,000 models to identify optimal hyperparameters (Figure 11.3). These are used in the performance estimation, which requires the fitting of another 500 models (5 folds * 100 repetitions; see Figure 11.3). To make the performance estimation processing chain even clearer, let us write down the commands we have given to the computer:

1. Performance level (upper left part of Figure 11.6): split the dataset into five spatially disjoint (outer) subfolds.
2. Tuning level (lower left part of Figure 11.6): use the first fold of the performance level and split it again spatially into five (inner) subfolds for the hyperparameter tuning. Use the 50 randomly selected hyperparameters in each of these inner subfolds, i.e., fit 250 models.
3. Performance estimation: Use the best hyperparameter combination from the previous step (tuning level) and apply it to the first outer fold in the performance level to estimate the performance (AUROC).
4. Repeat steps 2 and 3 for the remaining four outer folds.
5. Repeat steps 2 to 4, 100 times.

The process of hyperparameter tuning and performance estimation is computationally intensive. Model runtime can be reduced with parallelization, which can be done in a number of ways, depending on the operating system.

Before starting the parallelization, we ensure that the processing continues even if one of the models throws an error by setting `on.learner.error` to `warn`. This avoids the process stopping just because of one failed model, which is desirable on large model runs. To inspect the failed models once the processing is completed, we dump them:

```
configureMlr(on.learner.error = "warn", on.error.dump = TRUE)
```

To start the parallelization, we set the `mode` to `multicore` which will use `mclapply()` in the background on a single machine in the case of a Unix-based operating system.[10] Equivalently, `parallelStartSocket()` enables parallelization under Windows. `level` defines the level at which to enable parallelization, with `mlr.tuneParams` determining that the hyperparameter tuning level should be parallelized (see lower left part of Figure 11.6, `?parallelGetRegisteredLevels`, and the **mlr** parallelization tutorial[11] for details). We will use half of the available cores (set with the `cpus` parameter), a setting that allows possible other users to work on the same high performance computing cluster in case one is used (which was the case when we ran the code). Setting `mc.set.seed` to TRUE ensures that the randomly chosen hyperparameters during the tuning can be reproduced when running the code again. Unfortunately, `mc.set.seed` is only available under Unix-based systems.

```
library(parallelMap)
if (Sys.info()["sysname"] %in% c("Linux", "Darwin")) {
parallelStart(mode = "multicore",
               # parallelize the hyperparameter tuning level
               level = "mlr.tuneParams",
               # just use half of the available cores
               cpus = round(parallel::detectCores() / 2),
               mc.set.seed = TRUE)
}


if (Sys.info()["sysname"] == "Windows") {
  parallelStartSocket(level = "mlr.tuneParams",
                      cpus = round(parallel::detectCores() / 2))
}
```

Now we are set up for computing the nested spatial CV. Using a seed allows us to recreate the exact same spatial partitions when re-running the code. Specifying the `resample()` parameters follows the exact same procedure as

---

[10]See `?parallelStart` for further modes and github.com/berndbischl/parallelMap for more on the unified interface to popular parallelization back-ends.

[11]https://mlr-org.github.io/mlr-tutorial/release/html/parallelization/index.html#parallelization-levels

presented when using a GLM, the only difference being the `extract` argument. This allows the extraction of the hyperparameter tuning results which is important if we plan follow-up analyses on the tuning. After the processing, it is good practice to explicitly stop the parallelization with `parallelStop()`. Finally, we save the output object (`result`) to disk in case we would like to use it another R session. Before running the subsequent code, be aware that it is time-consuming: the 125,500 models took ~1/2hr on a server using 24 cores (see below).

```
set.seed(12345)
result = mlr::resample(learner = wrapped_lrn_ksvm,
                       task = task,
                       resampling = perf_level,
                       extract = getTuneResult,
                       measures = mlr::auc)
# stop parallelization
parallelStop()
# save your result, e.g.:
# saveRDS(result, "svm_sp_sp_rbf_50it.rds")
```

In case you do not want to run the code locally, we have saved a subset of the results[12] in the book's GitHub repo. They can be loaded as follows:

```
result = readRDS("extdata/spatial_cv_result.rds")
```

Note that runtime depends on many aspects: CPU speed, the selected algorithm, the selected number of cores and the dataset.

```
# Exploring the results
# runtime in minutes
round(result$runtime / 60, 2)
#> [1] 37.4
```

Even more important than the runtime is the final aggregated AUROC: the model's ability to discriminate the two classes.

```
# final aggregated AUROC
result$aggr
#> auc.test.mean
#>         0.758
# same as
```

---

[12]https://github.com/Robinlovelace/geocompr/blob/master/extdata/spatial_cv_result.rds

```
mean(result$measures.test$auc)
#> [1] 0.758
```

It appears that the GLM (aggregated AUROC was 0.78) is slightly better than the SVM in this specific case. However, using more than 50 iterations in the random search would probably yield hyperparameters that result in models with a better AUROC (Schratz et al., 2018). On the other hand, increasing the number of random search iterations would also increase the total number of models and thus runtime.

The estimated optimal hyperparameters for each fold at the performance estimation level can also be viewed. The following command shows the best hyperparameter combination of the first fold of the first iteration (recall this results from the first 5 * 50 model runs):

```
# winning hyperparameters of tuning step,
# i.e. the best combination out of 50 * 5 models
result$extract[[1]]$x
#> $C
#> [1] 0.458
#>
#> $sigma
#> [1] 0.023
```

The estimated hyperparameters have been used for the first fold in the first iteration of the performance estimation level which resulted in the following AUROC value:

```
result$measures.test[1, ]
#>    iter    auc
#> 1     1 0.799
```

So far spatial CV has been used to assess the ability of learning algorithms to generalize to unseen data. For spatial predictions, one would tune the hyperparameters on the complete dataset. This will be covered in Chapter 14.

11.6   Conclusions

Resampling methods are an important part of a data scientist's toolbox (James et al., 2013). This chapter used cross-validation to assess predictive performance of various models. As described in Section 11.4, observations with spatial coordinates may not be statistically independent due to spatial autocorrelation, violating a fundamental assumption of cross-validation. Spatial CV addresses this issue by reducing bias introduced by spatial autocorrelation.

The **mlr** package facilitates (spatial) resampling techniques in combination with the most popular statistical learning techniques including linear regression, semi-parametric models such as generalized additive models and machine learning techniques such as random forests, SVMs, and boosted regression trees (Bischl et al., 2016; Schratz et al., 2018). Machine learning algorithms often require hyperparameter inputs, the optimal 'tuning' of which can require thousands of model runs which require large computational resources, consuming much time, RAM and/or cores. **mlr** tackles this issue by enabling parallelization.

Machine learning overall, and its use to understand spatial data, is a large field and this chapter has provided the basics, but there is more to learn. We recommend the following resources in this direction:

- The **mlr** tutorials on Machine Learning in R[13] and Handling of spatial Data[14].
- An academic paper on hyperparameter tuning (Schratz et al., 2018).
- In case of spatio-temporal data, one should account for spatial and temporal autocorrelation when doing CV (Meyer et al., 2018).

11.7 Exercises

1. Compute the following terrain attributes from the `dem` datasets loaded with `data("landslides", package = "RSAGA")` with the help of R-GIS bridges (see Chapter 9):
   - Slope
   - Plan curvature
   - Profile curvature
   - Catchment area
2. Extract the values from the corresponding output rasters to the `landslides` data frame (`data(landslides, package = "RSAGA")` by adding new variables called `slope`, `cplan`, `cprof`, `elev` and `log_carea`. Keep all landslide initiation points and 175 randomly selected non-landslide points (see Section 11.2 for details).
3. Use the derived terrain attribute rasters in combination with a GLM to make a spatial prediction map similar to that shown in Figure

---

[13]https://mlr-org.github.io/mlr-tutorial/release/html/

[14]https://mlr-org.github.io/mlr-tutorial/release/html/handling_of_spatial_data/index.html

11.2. Running `data("study_mask", package = "spDataLarge")` attaches a mask of the study area.

4. Compute a 100-repeated 5-fold non-spatial cross-validation and spatial CV based on the GLM learner and compare the AUROC values from both resampling strategies with the help of boxplots (see Figure 11.5). Hint: You need to specify a non-spatial task and a non-spatial resampling strategy.

5. Model landslide susceptibility using a quadratic discriminant analysis (QDA, James et al., 2013). Assess the predictive performance (AUROC) of the QDA. What is the difference between the spatially cross-validated mean AUROC value of the QDA and the GLM? Hint: Before running the spatial cross-validation for both learners, set a seed to make sure that both use the same spatial partitions which in turn guarantees comparability.

6. Run the SVM without tuning the hyperparameters. Use the `rbfdot` kernel with $\sigma = 1$ and $C = 1$. Leaving the hyperparameters unspecified in **kernlab**'s `ksvm()` would otherwise initialize an automatic non-spatial hyperparameter tuning. For a discussion on the need for (spatial) tuning of hyperparameters, please refer to Schratz et al. (2018).

Part III

Applications

12

Transportation

Prerequisites

- This chapter uses the following packages:[1]

```
library(sf)
library(dplyr)
library(spDataLarge)
library(stplanr)        # geographic transport data package
library(tmap)           # visualization package (see Chapter 8)
```

12.1 Introduction

In few other sectors is geographic space more tangible than transport. The effort of moving (overcoming distance) is central to the 'first law' of geography, defined by Waldo Tobler in 1970 as follows (Miller, 2004):

> Everything is related to everything else, but near things are more related than distant things.

This 'law' is the basis for spatial autocorrelation and other key geographic concepts. It applies to phenomena as diverse as friendship networks and ecological diversity and can be explained by the costs of transport — in terms of time, energy and money — which constitute the 'friction of distance'. From

---

[1] **osmdata** and **nabor** must also be installed, although these packages do not need to be attached.

this perspective, transport technologies are disruptive, changing geographic relationships between geographic entities including mobile humans and goods: "the purpose of transportation is to overcome space" (Rodrigue et al., 2013).

Transport is an inherently geospatial activity. It involves traversing continuous geographic space between A and B, and infinite localities in between. It is therefore unsurprising that transport researchers have long turned to geocomputational methods to understand movement patterns and that transport problems are a motivator of geocomputational methods.

This chapter introduces the geographic analysis of transport systems at different geographic levels, including:

- **Areal units**: transport patterns can be understood with reference to zonal aggregates such as the main mode of travel (by car, bike or foot, for example) and average distance of trips made by people living in a particular zone, covered in Section 12.3.
- **Desire lines**: straight lines that represent 'origin-destination' data that records how many people travel (or could travel) between places (points or zones) in geographic space, the topic of Section 12.4.
- **Routes**: these are lines representing a path along the route network along the desire lines defined in the previous bullet point. We will see how to create them in Section 12.5.
- **Nodes**: these are points in the transport system that can represent common origins and destinations and public transport stations such as bus stops and rail stations, the topic of Section 12.6.
- **Route networks**: these represent the system of roads, paths and other linear features in an area and are covered in Section 12.7. They can be represented as geographic features (representing route segments) or structured as an interconnected graph, with the level of traffic on different segments referred to as 'flow' by transport modelers (Hollander, 2016).

Another key level is **agents**, mobile entities like you and me. These can be represented computationally thanks to software such as MATSim[2], which captures the dynamics of transport systems using an agent-based modeling (ABM) approach at high spatial and temporal resolution (Horni et al., 2016). ABM is a powerful approach to transport research with great potential for integration with R's spatial classes (Thiele, 2014; Lovelace and Dumont, 2016), but is outside the scope of this chapter. Beyond geographic levels and agents, the basic unit of analysis in most transport models is the **trip**, a single purpose journey from an origin 'A' to a destination 'B' (Hollander, 2016). Trips join-up the different levels of transport systems: they are usually represented as *desire lines* connecting *zone* centroids (*nodes*), they can be allocated onto the *route network* as *routes*, and are made by people who can be represented as *agents*.

---

[2] http://www.matsim.org/

Transport systems are dynamic systems adding additional complexity. The purpose of geographic transport modeling can be interpreted as simplifying this complexity in a way that captures the essence of transport problems. Selecting an appropriate level of geographic analysis can help simplify this complexity, to capture the essence of a transport system without losing its most important features and variables (Hollander, 2016).

Typically, models are designed to solve a particular problem. For this reason, this chapter is based around a policy scenario, introduced in the next section, that asks: how to increase cycling in the city of Bristol? Chapter 13 demonstrates another application of geocomputation: prioritising the location of new bike shops. There is a link between the chapters because bike shops may benefit from new cycling infrastructure, demonstrating an important feature of transport systems: they are closely linked to broader social, economic and land-use patterns.

12.2 A case study of Bristol

The case study used for this chapter is located in Bristol, a city in the west of England, around 30 km east of the Welsh capital Cardiff. An overview of the region's transport network is illustrated in Figure 12.1, which shows a diversity of transport infrastructure, for cycling, public transport, and private motor vehicles.

Bristol is the 10th largest city council in England, with a population of half a million people, although its travel catchment area is larger (see Section 12.3). It has a vibrant economy with aerospace, media, financial service and tourism companies, alongside two major universities. Bristol shows a high average income per capita but also contains areas of severe deprivation (Bristol City Council, 2015).

In terms of transport, Bristol is well served by rail and road links, and has a relatively high level of active travel. 19% of its citizens cycle and 88% walk at least once per month according to the Active People Survey[3] (the national average is 15% and 81%, respectively). 8% of the population said they cycled work in the 2011 census, compared with only 3% nationwide.

Despite impressive walking and cycling statistics, the city has a major congestion problem. Part of the solution is to continue to increase the proportion of trips made by cycling. Cycling has a greater potential to replace car trips than walking because of the speed of this mode, around 3-4 times faster than

---

[3] https://www.gov.uk/government/statistical-data-sets/how-often-and-time-spent-walking-and-cycling-at-local-authority-level-cw010#table-cw0103

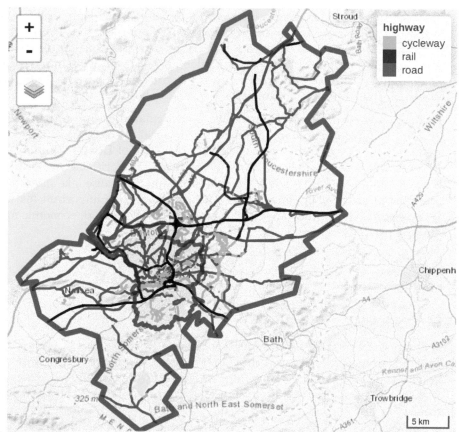

Leaflet | Tiles © Esri — Esri, DeLorme, NAVTEQ, TomTom, Intermap, iPC, USGS, FAO, NPS, NRCAN, GeoBase,
Kadaster NL, Ordnance Survey, Esri Japan, METI, Esri China (Hong Kong), and the GIS User Community

FIGURE 12.1: Bristol's transport network represented by colored lines for active (green), public (railways, black) and private motor (red) modes of travel. Blue border lines represent the inner city boundary and the larger Travel To Work Area (TTWA).

walking (with typical speeds[4] of 15-20 km/h vs 4-6 km/h for walking). There is an ambitious plan[5] to double the share of cycling by 2020.

In this policy context, the aim of this chapter, beyond demonstrating how geocomputation with R can be used to support sustainable transport planning, is to provide evidence for decision-makers in Bristol to decide how best to increase the share of walking and cycling in particular in the city. This high-level aim will be met via the following objectives:

---

[4]https://en.wikipedia.org/wiki/Bicycle_performance

[5]http://www.cyclingweekly.com/news/interview-bristols-mayor-george-ferguson-24114

- Describe the geographical pattern of transport behavior in the city.
- Identify key public transport nodes and routes along which cycling to rail stations could be encouraged, as the first stage in multi-model trips.
- Analyze travel 'desire lines', to find where many people drive short distances.
- Identify cycle route locations that will encourage less car driving and more cycling.

To get the wheels rolling on the practical aspects of this chapter, we begin by loading zonal data on travel patterns. These zone-level data are small but often vital for gaining a basic understanding of a settlement's overall transport system.

12.3 Transport zones

Although transport systems are primarily based on linear features and nodes — including pathways and stations — it often makes sense to start with areal data, to break continuous space into tangible units (Hollander, 2016). In addition to the boundary defining the study area (Bristol in this case), two zone types are of particular interest to transport researchers: and origin and destination zones. Often, the same geographic units are used for origins and destinations. However, different zoning systems, such as 'Workplace Zones[6]', may be appropriate to represent the increased density of trip destinations in areas with many 'trip attractors' such as schools and shops (Office for National Statistics, 2014).

The simplest way to define a study area is often the first matching boundary returned by OpenStreetMap, which can be obtained using **osmdata** with a command such as `bristol_region = osmdata::getbb("Bristol", format_out = "sf_polygon")`. This results in an `sf` object representing the bounds of the largest matching city region, either a rectangular polygon of the bounding box or a detailed polygonal boundary.[7] For Bristol, UK, a detailed polygon is returned, representing the official boundary of Bristol (see the inner blue boundary in Figure 12.1) but there are a couple of issues with this approach:

- The first OSM boundary returned by OSM may not be the official boundary used by local authorities.
- Even if OSM returns the official boundary, this may be inappropriate for transport research because they bear little relation to where people travel.

Travel to Work Areas (TTWAs) address these issues by creating a zoning system analogous to hydrological watersheds. TTWAs were first defined as

---

[6]https://data.gov.uk/dataset/workplace-zones-a-new-geography-for-workplace-statistics3

[7]In cases where the first match does not provide the right name, the country or region should be specified, for example `Bristol Tennessee` for a Bristol located in America.

contiguous zones within which 75% of the population travels to work (Coombes et al., 1986), and this is the definition used in this chapter. Because Bristol is a major employer attracting travel from surrounding towns, its TTWA is substantially larger than the city bounds (see Figure 12.1). The polygon representing this transport-orientated boundary is stored in the object `bristol_ttwa`, provided by the **spDataLarge** package loaded at the beginning of this chapter.

The origin and destination zones used in this chapter are the same: officially defined zones of intermediate geographic resolution (their official[8] name is Middle layer Super Output Areas or MSOAs). Each houses around 8,000 people. Such administrative zones can provide vital context to transport analysis, such as the type of people who might benefit most from particular interventions (e.g., Moreno-Monroy et al., 2017).

The geographic resolution of these zones is important: small zones with high geographic resolution are usually preferable but their high number in large regions can have consequences for processing (especially for origin-destination analysis in which the number of possibilities increases as a non-linear function of the number of zones) (Hollander, 2016).

Another issue with small zones is related to anonymity rules. To make it impossible to infer the identity of individuals in zones, detailed sociodemographic variables are often only available at low geographic resolution. Breakdowns of travel mode by age and sex, for example, are available at the Local Authority level in the UK, but not at the much higher Output Area level, each of which contains around 100 households. For further details, see www.ons.gov.uk/methodology/geography.

The 102 zones used in this chapter are stored in `bristol_zones`, as illustrated in Figure 12.2. Note the zones get smaller in densely populated areas: each houses a similar number of people. `bristol_zones` contains no attribute data on transport, however, only the name and code of each zone:

```
names(bristol_zones)
#> [1] "geo_code" "name"     "geometry"
```

To add travel data, we will undertake an *attribute join*, a common task described in Section 3.2.3. We will use travel data from the UK's 2011 census question on travel to work, data stored in `bristol_od`, which was provided by the ons.gov.uk[9] data portal. `bristol_od` is an origin-destination (OD) dataset on travel to work between zones from the UK's 2011 Census (see Section 12.4).

---

[8]https://www.ons.gov.uk/peoplepopulationandcommunity/populationandmigration/populationestimates/bulletins/annualsmallareapopulationestimates/2014-10-23

[9]https://www.ons.gov.uk/help/localstatistics

The first column is the ID of the zone of origin and the second column is the zone of destination. `bristol_od` has more rows than `bristol_zones`, representing travel *between* zones rather than the zones themselves:

```
nrow(bristol_od)
#> [1] 2910
nrow(bristol_zones)
#> [1] 102
```

The results of the previous code chunk shows that there are more than 10 OD pairs for every zone, meaning we will need to aggregate the origin-destination data before it is joined with `bristol_zones`, as illustrated below (origin-destination data is described in Section 12.4):

```
zones_attr = bristol_od %>%
  group_by(o) %>%
  summarize_if(is.numeric, sum) %>%
  dplyr::rename(geo_code = o)
```

The preceding chunk performed three main steps:

- Grouped the data by zone of origin (contained in the column o).
- Aggregated the variables in the `bristol_od` dataset *if* they were numeric, to find the total number of people living in each zone by mode of transport.[10]
- Renamed the grouping variable o so it matches the ID column `geo_code` in the `bristol_zones` object.

The resulting object `zones_attr` is a data frame with rows representing zones and an ID variable. We can verify that the IDs match those in the `zones` dataset using `%in%` operator as follows:

```
summary(zones_attr$geo_code %in% bristol_zones$geo_code)
#>    Mode    TRUE
#> logical     102
```

The results show that all 102 zones are present in the new object and that `zone_attr` is in a form that can be joined onto the zones.[11] This is done using the joining function `left_join()` (note that `inner_join()` would produce here the same result):

---

[10]the _if affix requires a TRUE/FALSE question to be asked of the variables, in this case 'is it numeric?' and only variables returning true are summarized.

[11]It would also be important to check that IDs match in the opposite direction on real data. This could be done by reversing the order of the ID's in the commend — summary(bristol_zones$geo_code %in% zones_attr$geo_code) — or by using setdiff() as follows: setdiff(bristol_zones$geo_code, zones_attr$geo_code).

```
zones_joined = left_join(bristol_zones, zones_attr, by = "geo_code")
sum(zones_joined$all)
#> [1] 238805
names(zones_joined)
#> [1] "geo_code"   "name"       "all"        "bicycle"    "foot"
#> [6] "car_driver" "train"      "geometry"
```

The result is `zones_joined`, which contains new columns representing the total number of trips originating in each zone in the study area (almost 1/4 of a million) and their mode of travel (by bicycle, foot, car and train). The geographic distribution of trip origins is illustrated in the left-hand map in Figure 12.2. This shows that most zones have between 0 and 4,000 trips originating from them in the study area. More trips are made by people living near the center of Bristol and fewer on the outskirts. Why is this? Remember that we are only dealing with trips within the study region: low trip numbers in the outskirts of the region can be explained by the fact that many people in these peripheral zones will travel to other regions outside of the study area. Trips outside the study region can be included in regional model by a special destination ID covering any trips that go to a zone not represented in the model (Hollander, 2016). The data in `bristol_od`, however, simply ignores such trips: it is an 'intra-zonal' model.

In the same way that OD datasets can be aggregated to the zone of origin, they can also be aggregated to provide information about destination zones. People tend to gravitate towards central places. This explains why the spatial distribution represented in the right panel in Figure 12.2 is relatively uneven, with the most common destination zones concentrated in Bristol city center. The result is `zones_od`, which contains a new column reporting the number of trip destinations by any mode, is created as follows:

```
zones_od = bristol_od %>%
  group_by(d) %>%
  summarize_if(is.numeric, sum) %>%
  dplyr::select(geo_code = d, all_dest = all) %>%
  inner_join(zones_joined, ., by = "geo_code")
```

A simplified version of Figure 12.2 is created with the code below (see `12-zones.R` in the `code`[12] folder of the book's GitHub repo to reproduce the figure and Section 8.2.6 for details on faceted maps with **tmap**):

```
qtm(zones_od, c("all", "all_dest")) +
  tm_layout(panel.labels = c("Origin", "Destination"))
```

---

[12]https://github.com/Robinlovelace/geocompr/tree/master/code

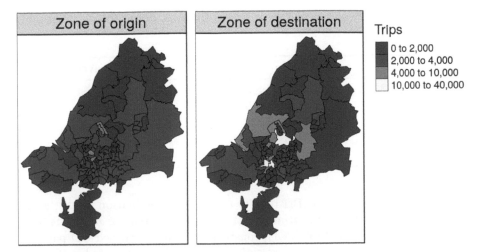

FIGURE 12.2: Number of trips (commuters) living and working in the region. The left map shows zone of origin of commute trips; the right map shows zone of destination (generated by the script 12-zones.R).

12.4 Desire lines

Unlike zones, which represent trip origins and destinations, desire lines connect the centroid of the origin and the destination zone, and thereby represent where people *desire* to go between zones. They represent the quickest 'bee line' or 'crow flies' route between A and B that would be taken, if it were not for obstacles such as buildings and windy roads getting in the way (we will see how to convert desire lines into routes in the next section).

We have already loaded data representing desire lines in the dataset `bristol_od`. This origin-destination (OD) data frame object represents the number of people traveling between the zone represented in `o` and `d`, as illustrated in Table 12.1. To arrange the OD data by all trips and then filter-out only the top 5, type (please refer to Chapter 3 for a detailed description of non-spatial attribute operations):

```
od_top5 = bristol_od %>%
  arrange(desc(all)) %>%
  top_n(5, wt = all)
```

The resulting table provides a snapshot of Bristolian travel patterns in terms of commuting (travel to work). It demonstrates that walking is the most popular mode of transport among the top 5 origin-destination pairs, that zone E02003043

TABLE 12.1: Sample of the origin-destination data stored in the object bristol od.

o	d	all	bicycle	foot	car_driver	train
E02003043	E02003043	1493	66	1296	64	8
E02003047	E02003043	1300	287	751	148	8
E02003031	E02003043	1221	305	600	176	7
E02003037	E02003043	1186	88	908	110	3
E02003034	E02003043	1177	281	711	100	7

is a popular destination (Bristol city center, the destination of all the top 5 OD pairs), and that the *intrazonal* trips, from one part of zone E02003043 to another (first row of Table 12.1), constitute the most traveled OD pair in the dataset. But from a policy perspective, the raw data presented in Table 12.1 is of limited use: aside from the fact that it contains only a tiny portion of the 2,910 OD pairs, it tells us little about *where* policy measures are needed, or *what proportion* of trips are made by walking and cycling. The following command calculates the percentage of each desire line that is made by these active modes:

```
bristol_od$Active = (bristol_od$bicycle + bristol_od$foot) /
   bristol_od$all * 100
```

There are two main types of OD pair: *interzonal* and *intrazonal*. Interzonal OD pairs represent travel between zones in which the destination is different from the origin. Intrazonal OD pairs represent travel within the same zone (see the top row of Table 12.1). The following code chunk splits od_bristol into these two types:

```
od_intra = filter(bristol_od, o == d)
od_inter = filter(bristol_od, o != d)
```

The next step is to convert the interzonal OD pairs into an sf object representing desire lines that can be plotted on a map with the **stplanr** function od2line().[13]

---

[13]od2line() works by matching the IDs in the first two columns of the bristol_od object to the zone_code ID column in the geographic zones_od object. Note that the operation emits a warning because od2line() works by allocating the start and end points of each origin-destination pair to the *centroid* of its zone of origin and destination. For real-world use one would use centroid values generated from projected data or, preferably, use *population-weighted* centroids (Lovelace et al., 2017).

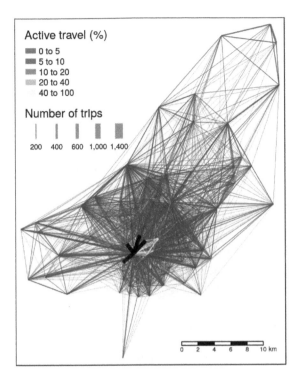

FIGURE 12.3: Desire lines representing trip patterns in Bristol, with width representing number of trips and color representing the percentage of trips made by active modes (walking and cycling). The four black lines represent the interzonal OD pairs in Table 7.1.

```
desire_lines = od2line(od_inter, zones_od)
```

An illustration of the results is presented in Figure 12.3, a simplified version of which is created with the following command (see the code in 12-desire.R to reproduce the figure exactly and Chapter 8 for details on visualization with **tmap**):

```
qtm(desire_lines, lines.lwd = "all")
```

The map shows that the city center dominates transport patterns in the region, suggesting policies should be prioritized there, although a number of peripheral sub-centers can also be seen. Next it would be interesting to have a look at the distribution of interzonal modes, e.g. between which zones is cycling the least or the most common means of transport.

12.5   Routes

From a geographer's perspective, routes are desire lines that are no longer straight: the origin and destination points are the same, but the pathway to get from A to B is more complex. Desire lines contain only two vertices (their beginning and end points) but routes can contain hundreds of vertices if they cover a large distance or represent travel patterns on an intricate road network (routes on simple grid-based road networks require relatively few vertices). Routes are generated from desire lines — or more commonly origin-destination pairs — using routing services which either run locally or remotely.

Local routing can be advantageous in terms of speed of execution and control over the weighting profile for different modes of transport. Disadvantages include the difficulty of representing complex networks locally; temporal dynamics (primarily due to traffic); and the need for specialized software such as 'pgRouting', an issue that developers of packages **stplanr** and **dodgr** seek to address.

Remote routing services, by contrast, use a web API to send queries about origins and destinations and return results generated on a powerful server running specialised software. This gives remote routing services various advantages, including that they usually:

- Update regularly.
- Have global coverage.
- Run on specialist hardware and software set-up for the job.

Disadvantages of remote routing services include speed (they rely on data transfer over the internet) and price (the Google routing API, for example, limits the number of free queries). The **googleway** package provides an interface to Google's routing API. Free (but rate limited) routing service include OSRM[14] and openrouteservice.org[15].

Instead of routing *all* desire lines generated in the previous section, which would be time and memory-consuming, we will focus on the desire lines of policy interest. The benefits of cycling trips are greatest when they replace car trips. Clearly, not all car trips can realistically be replaced by cycling. However, 5 km Euclidean distance (or around 6-8 km of route distance) can realistically be cycled by many people, especially if they are riding an electric bicycle ('ebike'). We will therefore only route desire lines along which a high (300+) number of car trips take place that are up to 5 km in distance. This routing is done by the **stplanr** function `line2route()` which takes straight lines

---

[14] http://project-osrm.org/
[15] https://openrouteservice.org/

in `Spatial` or `sf` objects, and returns 'bendy' lines representing routes on the transport network in the same class as the input.

```
desire_lines$distance = as.numeric(st_length(desire_lines))
desire_carshort = dplyr::filter(desire_lines, car_driver > 300 & distance < 5000)
```

```
route_carshort = line2route(desire_carshort, route_fun = route_osrm)
```

`st_length()` determines the length of a linestring, and falls into the distance relations category (see also Section 4.2.6). Subsequently, we apply a simple attribute filter operation (see Section 3.2.1) before letting the OSRM service do the routing on a remote server. Note that the routing only works with a working internet connection.

We could keep the new `route_carshort` object separate from the straight line representation of the same trip in `desire_carshort` but, from a data management perspective, it makes more sense to combine them: they represent the same trip. The new route dataset contains `distance` (referring to route distance this time) and `duration` fields (in seconds) which could be useful. However, for the purposes of this chapter, we are only interested in the geometry, from which route distance can be calculated. The following command makes use of the ability of simple features objects to contain multiple geographic columns:

```
desire_carshort$geom_car = st_geometry(route_carshort)
```

This allows plotting the desire lines along which many short car journeys take place alongside likely routes traveled by cars by referring to each geometry column separately (`desire_carshort$geometry` and `desire_carshort$geom_car` in this case). Making the width of the routes proportional to the number of car journeys that could potentially be replaced provides an effective way to prioritize interventions on the road network (Lovelace et al., 2017).

Plotting the results on an interactive map, with `mapview::mapview(desire_carshort$geom_car)` for example, shows that many short car trips take place in and around Bradley Stoke. It is easy to find explanations for the area's high level of car dependency: according to Wikipedia[16], Bradley Stoke is "Europe's largest new town built with private investment", suggesting limited public transport provision. Furthermore, the town is surrounded by large (cycling unfriendly) road structures, "such as junctions on both the M4 and M5 motorways" (Tallon, 2007).

There are many benefits of converting travel desire lines into likely routes of travel from a policy perspective, primary among them the ability to understand what it is about the surrounding environment that makes people travel by

---

[16] https://en.wikipedia.org/wiki/Bradley_Stoke

a particular mode. We discuss future directions of research building on the routes in Section 12.9. For the purposes of this case study, suffice to say that the roads along which these short car journeys travel should be prioritized for investigation to understand how they can be made more conducive to sustainable transport modes. One option would be to add new public transport nodes to the network. Such nodes are described in the next section.

12.6   Nodes

Nodes in geographic transport data are zero-dimensional features (points) among the predominantly one-dimensional features (lines) that comprise the network. There are two types of transport nodes:

1. Nodes not directly on the network such as zone centroids — covered in the next section — or individual origins and destinations such as houses and workplaces.
2. Nodes that are a part of transport networks, representing individual pathways, intersections between pathways (junctions) and points for entering or exiting a transport network such as bus stops and train stations.

Transport networks can be represented as graphs, in which each segment is connected (via edges representing geographic lines) to one or more other edges in the network. Nodes outside the network can be added with "centroid connectors", new route segments to nearby nodes on the network (Hollander, 2016).[17] Every node in the network is then connected by one or more 'edges' that represent individual segments on the network. We will see how transport networks can be represented as graphs in Section 12.7.

Public transport stops are particularly important nodes that can be represented as either type of node: a bus stop that is part of a road, or a large rail station that is represented by its pedestrian entry point hundreds of meters from railway tracks. We will use railway stations to illustrate public transport nodes, in relation to the research question of increasing cycling in Bristol. These stations are provided by **spDataLarge** in `bristol_stations`.

A common barrier preventing people from switching away from cars for commuting to work is that the distance from home to work is too far to walk or cycle. Public transport can reduce this barrier by providing a fast and high-volume option for common routes into cities. From an active travel perspective, public transport 'legs' of longer journeys divide trips into three:

---

[17]The location of these connectors should be chosen carefully because they can lead to over-estimates of traffic volumes in their immediate surroundings (Jafari et al., 2015).

- The origin leg, typically from residential areas to public transport stations.
- The public transport leg, which typically goes from the station nearest a trip's origin to the station nearest its destination.
- The destination leg, from the station of alighting to the destination.

Building on the analysis conducted in Section 12.4, public transport nodes can be used to construct three-part desire lines for trips that can be taken by bus and (the mode used in this example) rail. The first stage is to identify the desire lines with most public transport travel, which in our case is easy because our previously created dataset desire_lines already contains a variable describing the number of trips by train (the public transport potential could also be estimated using public transport routing services such as OpenTripPlanner[18]). To make the approach easier to follow, we will select only the top three desire lines in terms of rails use:

```
desire_rail = top_n(desire_lines, n = 3, wt = train)
```

The challenge now is to 'break-up' each of these lines into three pieces, representing travel via public transport nodes. This can be done by converting a desire line into a multiline object consisting of three line geometries representing origin, public transport and destination legs of the trip. This operation can be divided into three stages: matrix creation (of origins, destinations and the 'via' points representing rail stations), identification of nearest neighbors and conversion to multilines. These are undertaken by line_via(). This **stplanr** function takes input lines and points and returns a copy of the desire lines — see the Desire Lines Extended[19] vignette on the geocompr.github.io website and ?line_via for details on how this works. The output is the same as the input line, except it has new geometry columns representing the journey via public transport nodes, as demonstrated below:

```
ncol(desire_rail)
#> [1] 10
desire_rail = line_via(desire_rail, bristol_stations)
ncol(desire_rail)
#> [1] 13
```

As illustrated in Figure 12.4, the initial desire_rail lines now have three additional geometry list columns representing travel from home to the origin station, from there to the destination, and finally from the destination station to the destination. In this case, the destination leg is very short (walking distance) but the origin legs may be sufficiently far to justify investment in cycling infrastructure to encourage people to cycle to the stations on the

---

[18] http://www.opentripplanner.org/

[19] https://geocompr.github.io/geocompkg/articles/linevia.html

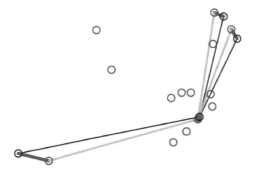

FIGURE 12.4: Station nodes (red dots) used as intermediary points that convert straight desire lines with high rail usage (black) into three legs: to the origin station (red) via public transport (gray) and to the destination (a very short blue line).

outward leg of peoples' journey to work in the residential areas surrounding the three origin stations in Figure 12.4.

12.7 Route networks

The data used in this section was downloaded using **osmdata**. To avoid having to request the data from OSM repeatedly, we will use the `bristol_ways` object, which contains point and line data for the case study area (see `?bristol_ways`):

```
summary(bristol_ways)
#>      highway        maxspeed          ref              geometry
#>  cycleway:1317   30 mph : 925   A38     : 214   LINESTRING   :4915
#>  rail    : 832   20 mph : 556   A432    : 146   epsg:4326    :    0
#>  road    :2766   40 mph : 397   M5      : 144   +proj=long...:    0
#>                  70 mph : 328   A4018   : 124
#>                  50 mph : 158   A420    : 115
#>                  (Other): 490   (Other):1877
#>                  NA's   :2061   NA's    :2295
```

The above code chunk loaded a simple feature object representing around 3,000 segments on the transport network. This an easily manageable dataset size (transport datasets can be large, but it's best to start small).

As mentioned, route networks can usefully be represented as mathematical graphs, with nodes on the network connected by edges. A number of R packages have been developed for dealing with such graphs, notably **igraph**. One can manually convert a route network into an `igraph` object, but the geographic attributes will be lost. To overcome this issue `SpatialLinesNetwork()` was developed in the **stplanr** package to represent route networks simultaneously as graphs *and* a set of geographic lines. This function is demonstrated below using a subset of the `bristol_ways` object used in previous sections.

```
ways_freeway = bristol_ways %>% filter(maxspeed == "70 mph")
ways_sln = SpatialLinesNetwork(ways_freeway)
slotNames(ways_sln)
#> [1] "sl"          "g"          "nb"          "weightfield"
weightfield(ways_sln)
#> [1] "length"
class(ways_sln@g)
#> [1] "igraph"
```

The output of the previous code chunk shows that `ways_sln` is a composite object with various 'slots'. These include: the spatial component of the network (named `sl`), the graph component (`g`) and the 'weightfield', the edge variable used for shortest path calculation (by default segment distance). `ways_sln` is of class `sfNetwork`, defined by the S4 class system. This means that each component can be accessed using the `@` operator, which is used below to extract its graph component and process it using the **igraph** package, before plotting the results in geographic space. In the example below, the 'edge betweenness', meaning the number of shortest paths passing through each edge, is calculated (see `?igraph::betweenness` for further details and Figure 12.5). The results demonstrate that each graph edge represents a segment: the segments near the center of the road network have the greatest betweenness scores.

```
g = ways_sln@g
e = igraph::edge_betweenness(ways_sln@g)
plot(ways_sln@sl$geometry, lwd = e / 500)
```

One can also find the shortest route between origins and destinations using this graph representation of the route network. This can be done with functions such as `sum_network_routes()` from **stplanr**, which undertakes 'local routing' (see Section 12.5).

12.8 Prioritizing new infrastructure

This chapter's final practical section demonstrates the policy-relevance of geocomputation for transport applications by identifying locations where new

FIGURE 12.5: Illustration of a small route network, with segment thickness proportional to its betweenless, generated using the igraph package and described in the text.

transport infrastructure may be needed. Clearly, the types of analysis presented here would need to be extended and complemented by other methods to be used in real-world applications, as discussed in Section 12.9. However, each stage could be useful on its own, and feed into wider analyses. To summarize, these were: identifying short but car-dependent commuting routes (generated from desire lines) in Section 12.5; creating desire lines representing trips to rail stations in Section 12.6; and analysis of transport systems at the route network using graph theory in Section 12.7.

The final code chunk of this chapter combines these strands of analysis. It adds the car-dependent routes in route_carshort with a newly created object, route_rail and creates a new column representing the amount of travel along the centroid-to-centroid desire lines they represent:

```
route_rail = desire_rail %>%
  st_set_geometry("leg_orig") %>%
  line2route(route_fun = route_osrm) %>%
  st_set_crs(4326)
```

```
route_cycleway = rbind(route_rail, route_carshort)
route_cycleway$all = c(desire_rail$all, desire_carshort$all)
```

The results of the preceding code are visualized in Figure 12.6, which shows routes with high levels of car dependency and highlights opportunities for cycling rail stations (the subsequent code chunk creates a simple version of the

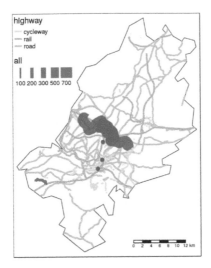

FIGURE 12.6: Potential routes along which to prioritise cycle infrastructure in Bristol, based on access key rail stations (red dots) and routes with many short car journeys (north of Bristol surrounding Stoke Bradley). Line thickness is proportional to number of trips.

figure — see `code/12-cycleways.R` to reproduce the figure exactly). The method has some limitations: in reality, people do not travel to zone centroids or always use the shortest route algorithm for a particular mode. However, the results demonstrate routes along which cycle paths could be prioritized from car dependency and public transport perspectives.

```
qtm(route_cycleway, lines.lwd = "all")
```

The results may look more attractive in an interactive map, but what do they mean? The routes highlighted in Figure 12.6 suggest that transport systems are intimately linked to the wider economic and social context. The example of Stoke Bradley is a case in point: its location, lack of public transport services and active travel infrastructure help explain why it is so highly car-dependent. The wider point is that car dependency has a spatial distribution which has implications for sustainable transport policies (Hickman et al., 2011).

12.9 Future directions of travel

This chapter provides a taste of the possibilities of using geocomputation for transport research. It has explored some key geographic elements that make-up a city's transport system using open data and reproducible code. The results could help plan where investment is needed.

Transport systems operate at multiple interacting levels, meaning that geo-computational methods have great potential to generate insights into how they work. There is much more that could be done in this area: it would be possible to build on the foundations presented in this chapter in many directions. Transport is the fastest growing source of greenhouse gas emissions in many countries, and is set to become "the largest GHG emitting sector, especially in developed countries" (see EURACTIV.com[20]). Because of the highly unequal distribution of transport-related emissions across society, and the fact that transport (unlike food and heating) is not essential for well-being, there is great potential for the sector to rapidly decarbonize through demand reduction, electrification of the vehicle fleet and the uptake of active travel modes such as walking and cycling. Further exploration of such 'transport futures' at the local level represents promising direction of travel for transport-related geocomputational research.

Methodologically, the foundations presented in this chapter could be extended by including more variables in the analysis. Characteristics of the route such as speed limits, busyness and the provision of protected cycling and walking paths could be linked to 'mode-split' (the proportion of trips made by different modes of transport). By aggregating OpenStreetMap data using buffers and geographic data methods presented in Chapters 3 and 4, for example, it would be possible to detect the presence of green space in close proximity to transport routes. Using R's statistical modeling capabilities, this could then be used to predict current and future levels of cycling, for example.

This type of analysis underlies the Propensity to Cycle Tool (PCT), a publicly accessible (see www.pct.bike[21]) mapping tool developed in R that is being used to prioritize investment in cycling across England (Lovelace et al., 2017). Similar tools could be used to encourage evidence-based transport policies related to other topics such as air pollution and public transport access around the world.

12.10 Exercises

1. What is the total distance of cycleways that would be constructed if all the routes presented in Figure 12.6 were to be constructed?
   - Bonus: find two ways of arriving at the same answer.

---

[20]https://www.euractiv.com/section/agriculture-food/opinion/transport-needs-to-do-a-lot-more-to-fight-climate-change/

[21]http://www.pct.bike/

2. What proportion of trips represented in the `desire_lines` are accounted for in the `route_cycleway` object?
   - Bonus: what proportion of trips cross the proposed routes?
   - Advanced: write code that would increase this proportion.
3. The analysis presented in this chapter is designed for teaching how geocomputation methods can be applied to transport research. If you were to do this 'for real' for local government or a transport consultancy, what top 3 things would you do differently?
4. Clearly, the routes identified in Figure 12.6 only provide part of the picture. How would you extend the analysis to incorporate more trips that could potentially be cycled?
5. Imagine that you want to extend the scenario by creating key *areas* (not routes) for investment in place-based cycling policies such as car-free zones, cycle parking points and reduced car parking strategy. How could raster data assist with this work?
   - Bonus: develop a raster layer that divides the Bristol region into 100 cells (10 by 10) and provide a metric related to transport policy, such as number of people trips that pass through each cell by walking or the average speed limit of roads, from the `bristol_ways` dataset (the approach taken in Chapter 13).

13

Geomarketing

Prerequisites

- This chapter requires the following packages (**revgeo** must also be installed):

```
library(sf)
library(dplyr)
library(purrr)
library(raster)
library(osmdata)
library(spDataLarge)
```

- Required data will be downloaded in due course. As a convenience to the reader and to ensure easy reproducibility, we have made available the downloaded data in the **spDataLarge** package.

13.1 Introduction

This chapter demonstrates how the skills learned in Parts I and II can be applied to a particular domain: geomarketing (sometimes also referred to as location analysis or location intelligence). This is a broad field of research and commercial application. A typical example is where to locate a new shop. The aim here is to attract most visitors and, ultimately, make the most profit. There are also many non-commercial applications that can use the technique for public benefit, for example where to locate new health services (Tomintz et al., 2008).

People are fundamental to location analysis, in particular where they are likely to spend their time and other resources. Interestingly, ecological concepts and models are quite similar to those used for store location analysis. Animals and plants can best meet their needs in certain 'optimal' locations, based on variables that change over space [Muenchow et al. (2018); see also chapter 14].

This is one of the great strengths of geocomputation and GIScience in general. Concepts and methods are transferable to other fields. Polar bears, for example, prefer northern latitudes where temperatures are lower and food (seals and sea lions) is plentiful. Similarly, humans tend to congregate in certain places, creating economic niches (and high land prices) analogous to the ecological niche of the Arctic. The main task of location analysis is to find out where such 'optimal locations' are for specific services, based on available data. Typical research questions include:

- Where do target groups live and which areas do they frequent?
- Where are competing stores or services located?
- How many people can easily reach specific stores?
- Do existing services over- or under-exploit the market potential?
- What is the market share of a company in a specific area?

This chapter demonstrates how geocomputation can answer such questions based on a hypothetical case study based on real data.

13.2   Case study: bike shops in Germany

Imagine you are starting a chain of bike shops in Germany. The stores should be placed in urban areas with as many potential customers as possible. Additionally, a hypothetical survey (invented for this chapter, not for commercial use!) suggests that single young males (aged 20 to 40) are most likely to buy your products: this is the *target audience*. You are in the lucky position to have sufficient capital to open a number of shops. But where should they be placed? Consulting companies (employing geomarketing analysts) would happily charge high rates to answer such questions. Luckily, we can do so ourselves with the help of open data and open source software. The following sections will demonstrate how the techniques learned during the first chapters of the book can be applied to undertake the following steps:

- Tidy the input data from the German census (Section 13.3).
- Convert the tabulated census data into raster objects (Section 13.4).
- Identify metropolitan areas with high population densities (Section 13.5).
- Download detailed geographic data (from OpenStreetMap, with **osmdata**) for these areas (Section 13.6).
- Create rasters for scoring the relative desirability of different locations using map algebra (Section 13.7).

Although we have applied these steps to a specific case study, they could be generalized to many scenarios of store location or public service provision.

13.3 Tidy the input data

The German government provides gridded census data at either 1 km or 100 m resolution. The following code chunk downloads, unzips and reads in the 1 km data.

```
download.file("https://tinyurl.com/ybtpkwxz",
              destfile = "census.zip", mode = "wb")
unzip("census.zip") # unzip the files
census_de = readr::read_csv2(list.files(pattern = "Gitter.csv"))
```

As a convenience to the reader, the corresponding data has been put into **spDataLarge** and can be accessed as follows

```
data("census_de", package = "spDataLarge")
```

The `census_de` object is a data frame containing 13 variables for more than 300,000 grid cells across Germany. For our work, we only need a subset of these: Easting (`x`) and Northing (`y`), number of inhabitants (population; `pop`), mean average age (`mean_age`), proportion of women (`women`) and average household size (`hh_size`). These variables are selected and renamed from German into English in the code chunk below and summarized in Table 13.1. Further, `mutate_all()` is used to convert values -1 and -9 (meaning unknown) to `NA`.

```
# pop = population, hh_size = household size
input = dplyr::select(census_de, x = x_mp_1km, y = y_mp_1km, pop = Einwohner,
                      women = Frauen_A, mean_age = Alter_D,
                      hh_size = HHGroesse_D)
# set -1 and -9 to NA
input_tidy = mutate_all(input, list(~ifelse(. %in% c(-1, -9), NA, .)))
```

13.4 Create census rasters

After the preprocessing, the data can be converted into a raster stack or brick (see Sections 2.3.3 and 3.3.1). `rasterFromXYZ()` makes this really easy. It requires an input data frame where the first two columns represent coordinates on a regular grid. All the remaining columns (here: `pop`, `women`, `mean_age`, `hh_size`) will serve as input for the raster brick layers (Figure 13.1; see also `code/13-location-jm.R` in our github repository).

TABLE 13.1: Categories for each variable in census data from Daten-satzbeschreibung...xlsx located in the downloaded file census.zip (see Figure 13.1 for their spatial distribution).

class	Population	% female	Mean age	Household size
1	3-250	0-40	0-40	1-2
2	250-500	40-47	40-42	2-2.5
3	500-2000	47-53	42-44	2.5-3
4	2000-4000	53-60	44-47	3-3.5
5	4000-8000	>60	>47	>3.5
6	>8000			

```
input_ras = rasterFromXYZ(input_tidy, crs = st_crs(3035)$proj4string)
```

```
input_ras
#> class : RasterBrick
#> dimensions : 868, 642, 557256, 4 (nrow, ncol, ncell, nlayers)
#> resolution : 1000, 1000 (x, y)
#> extent : 4031000, 4673000, 2684000, 3552000 (xmin, xmax, ymin, ymax)
#> coord. ref. : +proj=laea +lat_0=52 +lon_0=10
#> names      :  pop, women, mean_age, hh_size
#> min values :  1,     1,        1,       1
#> max values :  6,     5,        5,       5
```

Note that we are using an equal-area projection (EPSG:3035; Lambert Equal Area Europe), i.e., a projected CRS where each grid cell has the same area, here 1000 x 1000 square meters. Since we are using mainly densities such as the number of inhabitants or the portion of women per grid cell, it is of utmost importance that the area of each grid cell is the same to avoid 'comparing apples and oranges'. Be careful with geographic CRS where grid cell areas constantly decrease in poleward directions (see also Section 2.4 and Chapter 6).

The next stage is to reclassify the values of the rasters stored in input_ras in accordance with the survey mentioned in Section 13.2, using the **raster** function reclassify(), which was introduced in Section 4.3.3. In the case of the population data, we convert the classes into a numeric data type using class means. Raster cells are assumed to have a population of 127 if they have a value of 1 (cells in 'class 1' contain between 3 and 250 inhabitants) and 375 if they have a value of 2 (containing 250 to 500 inhabitants), and so on (see Table

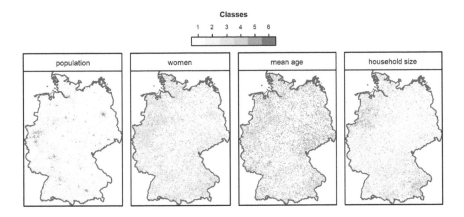

FIGURE 13.1: Gridded German census data of 2011 (see Table 13.1 for a description of the classes).

13.1). A cell value of 8000 inhabitants was chosen for 'class 6' because these cells contain more than 8000 people. Of course, these are approximations of the true population, not precise values.[1] However, the level of detail is sufficient to delineate metropolitan areas (see next section).

In contrast to the pop variable, representing absolute estimates of the total population, the remaining variables were re-classified as weights corresponding with weights used in the survey. Class 1 in the variable women, for instance, represents areas in which 0 to 40% of the population is female; these are reclassified with a comparatively high weight of 3 because the target demographic is predominantly male. Similarly, the classes containing the youngest people and highest proportion of single households are reclassified to have high weights.

```
rcl_pop = matrix(c(1, 1, 127, 2, 2, 375, 3, 3, 1250,
                   4, 4, 3000, 5, 5, 6000, 6, 6, 8000),
                 ncol = 3, byrow = TRUE)
rcl_women = matrix(c(1, 1, 3, 2, 2, 2, 3, 3, 1, 4, 5, 0),
                   ncol = 3, byrow = TRUE)
rcl_age = matrix(c(1, 1, 3, 2, 2, 0, 3, 5, 0),
                 ncol = 3, byrow = TRUE)
```

---

[1]The potential error introduced during this reclassification stage will be explored in the exercises.

```
rcl_hh = rcl_women
rcl = list(rcl_pop, rcl_women, rcl_age, rcl_hh)
```

Note that we have made sure that the order of the reclassification matrices in the list is the same as for the elements of `input_ras`. For instance, the first element corresponds in both cases to the population. Subsequently, the `for`-loop applies the reclassification matrix to the corresponding raster layer. Finally, the code chunk below ensures the `reclass` layers have the same name as the layers of `input_ras`.

```
reclass = input_ras
for (i in seq_len(nlayers(reclass))) {
  reclass[[i]] = reclassify(x = reclass[[i]], rcl = rcl[[i]], right = NA)
}
names(reclass) = names(input_ras)
```

```
reclass
#> ... (full output not shown)
#> names        :  pop, women, mean_age, hh_size
#> min values   :  127,     0,        0,       0
#> max values   : 8000,     3,        3,       3
```

13.5   Define metropolitan areas

We define metropolitan areas as pixels of 20 km^2 inhabited by more than 500,000 people. Pixels at this coarse resolution can rapidly be created using `aggregate()`, as introduced in Section 5.3.3. The command below uses the argument `fact = 20` to reduce the resolution of the result twenty-fold (recall the original raster resolution was 1 km^2):

```
pop_agg = aggregate(reclass$pop, fact = 20, fun = sum)
```

The next stage is to keep only cells with more than half a million people.

```
pop_agg = pop_agg[pop_agg > 500000, drop = FALSE]
```

Plotting this reveals eight metropolitan regions (Figure 13.2). Each region consists of one or more raster cells. It would be nice if we could join all cells belonging to one region. **raster**'s `clump()` command does exactly that.

Subsequently, `rasterToPolygons()` converts the raster object into spatial polygons, and `st_as_sf()` converts it into an `sf`-object.

```
polys = pop_agg %>%
  clump() %>%
  rasterToPolygons() %>%
  st_as_sf()
```

`polys` now features a column named `clumps` which indicates to which metropolitan region each polygon belongs and which we will use to dissolve the polygons into coherent single regions (see also Section 5.2.6):

```
metros = polys %>%
  group_by(clumps) %>%
  summarize()
```

Given no other column as input, `summarize()` only dissolves the geometry.

The resulting eight metropolitan areas suitable for bike shops (Figure 13.2; see also `code/13-location-jm.R` for creating the figure) are still missing a name. A reverse geocoding approach can settle this problem. Given a coordinate, reverse geocoding finds the corresponding address. Consequently, extracting the centroid coordinate of each metropolitan area can serve as an input for a reverse A API. The **revgeo** package provides access to the open source Photon geocoder for OpenStreetMap, Google Maps and Bing. By default, it uses the Photon API. `revgeo::revgeo()` only accepts geographical coordinates (latitude/longitude); therefore, the first requirement is to bring the metropolitan polygons into an appropriate coordinate reference system (Chapter 6).

```
metros_wgs = st_transform(metros, 4326)
coords = st_centroid(metros_wgs) %>%
  st_coordinates() %>%
  round(4)
```

Choosing `frame` as `revgeocode()`'s `output` option will give back a `data.frame` with several columns referring to the location including the street name, house number and city.

```
library(revgeo)
metro_names = revgeo(longitude = coords[, 1], latitude = coords[, 2],
                     output = "frame")
```

To make sure that the reader uses the exact same results, we have put them into **spDataLarge**:

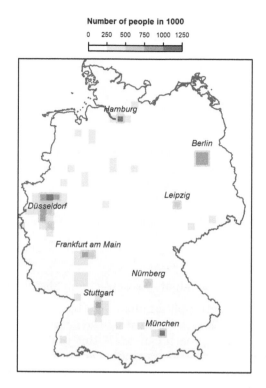

FIGURE 13.2: The aggregated population raster (resolution: 20 km) with the identified metropolitan areas (golden polygons) and the corresponding names.

```
# attach metro_names from spDataLarge
data("metro_names", package = "spDataLarge")
```

Overall, we are satisfied with the city column serving as metropolitan names (Table 13.2) apart from one exception, namely Wülfrath which belongs to the greater region of Düsseldorf. Hence, we replace Wülfrath with Düsseldorf (Figure 13.2). Umlauts like ü might lead to trouble further on, for example when determining the bounding box of a metropolitan area with opq() (see further below), which is why we avoid them.

```
metro_names = dplyr::pull(metro_names, city) %>%
  as.character() %>%
  ifelse(. == "Wülfrath", "Duesseldorf", .)
```

TABLE 13.2: Result of the reverse geocoding.

city	state
Hamburg	Hamburg
Berlin	Berlin
Wülfrath	North Rhine-Westphalia
Leipzig	Saxony
Frankfurt am Main	Hesse
Nuremberg	Bavaria
Stuttgart	Baden-Württemberg
Munich	Bavaria

13.6 Points of interest

The **osmdata** package provides easy-to-use access to OSM data (see also Section 7.2). Instead of downloading shops for the whole of Germany, we restrict the query to the defined metropolitan areas, reducing computational load and providing shop locations only in areas of interest. The subsequent code chunk does this using a number of functions including:

- `map()` (the **tidyverse** equivalent of `lapply()`), which iterates through all eight metropolitan names which subsequently define the bounding box in the OSM query function `opq()` (see Section 7.2).
- `add_osm_feature()` to specify OSM elements with a key value of `shop` (see wiki.openstreetmap.org[2] for a list of common key:value pairs).
- `osmdata_sf()`, which converts the OSM data into spatial objects (of class `sf`).
- `while()`, which tries repeatedly (three times in this case) to download the data if it fails the first time.[3] Before running this code: please consider it will download almost 2GB of data. To save time and resources, we have output into **spDataLarge** and should already be available in your environment as an object called `shops`.

```
shops = map(metro_names, function(x) {
  message("Downloading shops of: ", x, "\n")
  # give the server a bit time
  Sys.sleep(sample(seq(5, 10, 0.1), 1))
  query = opq(x) %>%
    add_osm_feature(key = "shop")
```

---

[2]http://wiki.openstreetmap.org/wiki/Map_Features
[3]The OSM-download will sometimes fail at the first attempt.

```
points = osmdata_sf(query)
# request the same data again if nothing has been downloaded
iter = 2
while (nrow(points$osm_points) == 0 & iter > 0) {
  points = osmdata_sf(query)
  iter = iter - 1
}
points = st_set_crs(points$osm_points, 4326)
})
```

It is highly unlikely that there are no shops in any of our defined metropolitan areas. The following `if` condition simply checks if there is at least one shop for each region. If not, we would try to download the shops again for this/these specific region/s.

```
# checking if we have downloaded shops for each metropolitan area
ind = map(shops, nrow) == 0
if (any(ind)) {
  message("There are/is still (a) metropolitan area/s without any features:\n",
          paste(metro_names[ind], collapse = ", "), "\nPlease fix it!")
}
```

To make sure that each list element (an `sf` data frame) comes with the same columns, we only keep the `osm_id` and the `shop` columns with the help of another `map` loop. This is not a given since OSM contributors are not equally meticulous when collecting data. Finally, we `rbind` all shops into one large `sf` object.

```
# select only specific columns
shops = map(shops, dplyr::select, osm_id, shop)
# putting all list elements into a single data frame
shops = do.call(rbind, shops)
```

It would have been easier to simply use `map_dfr()`. Unfortunately, so far it does not work in harmony with `sf` objects. Please note that the `shops` object is also available in the `spDataLarge` package:

```
data("shops", package = "spDataLarge")
```

The only thing left to do is to convert the spatial point object into a raster (see Section 5.4.3). The `sf` object, `shops`, is converted into a raster having the same parameters (dimensions, resolution, CRS) as the `reclass` object. Importantly, the `count()` function is used here to calculate the number of shops in each cell.

If the shop column were used instead of the osm_id column, we would have retrieved fewer shops per grid cell. This is because the shop column contains NA values, which the count() function omits when rasterizing vector objects.

The result of the subsequent code chunk is therefore an estimate of shop density (shops/km^2). st_transform() is used before rasterize() to ensure the CRS of both inputs match.

```
shops = st_transform(shops, proj4string(reclass))
# create poi raster
poi = rasterize(x = shops, y = reclass, field = "osm_id", fun = "count")
```

As with the other raster layers (population, women, mean age, household size) the poi raster is reclassified into four classes (see Section 13.4). Defining class intervals is an arbitrary undertaking to a certain degree. One can use equal breaks, quantile breaks, fixed values or others. Here, we choose the Fisher-Jenks natural breaks approach which minimizes within-class variance, the result of which provides an input for the reclassification matrix.

```
# construct reclassification matrix
int = classInt::classIntervals(values(poi), n = 4, style = "fisher")
int = round(int$brks)
rcl_poi = matrix(c(int[1], rep(int[-c(1, length(int))], each = 2),
                   int[length(int)] + 1), ncol = 2, byrow = TRUE)
rcl_poi = cbind(rcl_poi, 0:3)
# reclassify
poi = reclassify(poi, rcl = rcl_poi, right = NA)
names(poi) = "poi"
```

13.7 Identifying suitable locations

The only steps that remain before combining all the layers are to add POI and delete the population from poi to the reclass raster stack and remove the population layer from it.. The reasoning for the latter is twofold. First of all, we have already delineated metropolitan areas, that is areas where the population density is above average compared to the rest of Germany. Second, though it is advantageous to have many potential customers within a specific catchment area, the sheer number alone might not actually represent the desired target group. For instance, residential tower blocks are areas with a high population density but not necessarily with a high purchasing power for expensive cycle

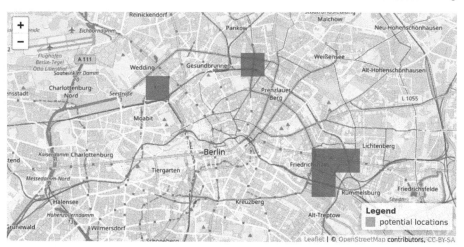

FIGURE 13.3: Suitable areas (i.e., raster cells with a score > 9) in accordance with our hypothetical survey for bike stores in Berlin.

components. This is achieved with the complementary functions `addLayer()` and `dropLayer()`:

```
# add poi raster
reclass = addLayer(reclass, poi)
# delete population raster
reclass = dropLayer(reclass, "pop")
```

In common with other data science projects, data retrieval and 'tidying' have consumed much of the overall workload so far. With clean data, the final step — calculating a final score by summing all raster layers — can be accomplished in a single line of code.

```
# calculate the total score
result = sum(reclass)
```

For instance, a score greater than 9 might be a suitable threshold indicating raster cells where a bike shop could be placed (Figure 13.3; see also `code/13-location-jm.R`).

13.8 Discussion and next steps

The presented approach is a typical example of the normative usage of a GIS (Longley, 2015). We combined survey data with expert-based knowledge and assumptions (definition of metropolitan areas, defining class intervals, definition of a final score threshold). It should be clear that this approach is not suitable for scientific knowledge advancement but is a very applied way of information extraction. This is to say, we can only suspect based on common sense that we have identified areas suitable for bike shops. However, we have no proof that this is in fact the case.

A few other things remained unconsidered but might improve the analysis:

- We used equal weights when calculating the final scores. But is, for example, the household size as important as the portion of women or the mean age?
- We used all points of interest. Maybe it would be wiser to use only those which might be interesting for bike shops such as do-it-yourself, hardware, bicycle, fishing, hunting, motorcycles, outdoor and sports shops (see the range of shop values available on the OSM Wiki[4]).
- Data at a better resolution may change and improve the output. For example, there is also population data at a finer resolution (100 m; see exercises).
- We have used only a limited set of variables. For example, the INSPIRE geoportal[5] might contain much more data of possible interest to our analysis (see also Section 7.2). The bike paths density might be another interesting variable as well as the purchasing power or even better the retail purchasing power for bikes.
- Interactions remained unconsidered, such as a possible interaction between the portion of men and single households. However, to find out about such an interaction we would need customer data.

In short, the presented analysis is far from perfect. Nevertheless, it should have given you a first impression and understanding of how to obtain and deal with spatial data in R within a geomarketing context.

Finally, we have to point out that the presented analysis would be merely the first step of finding suitable locations. So far we have identified areas, 1 by 1 km in size, potentially suitable for a bike shop in accordance with our survey. We could continue the analysis as follows:

- Find an optimal location based on number of inhabitants within a specific catchment area. For example, the shop should be reachable for as many people as possible within 15 minutes of traveling bike distance (catchment area routing). Thereby, we should account for the fact that the further away

---

[4]http://wiki.openstreetmap.org/wiki/Map_Features#Shop

[5]http://inspire-geoportal.ec.europa.eu/discovery/

the people are from the shop, the more unlikely it becomes that they actually visit it (distance decay function).

- Also it would be a good idea to take into account competitors. That is, if there already is a bike shop in the vicinity of the chosen location, one has to distribute possible customers (or sales potential) between the competitors (Huff, 1963; Wieland, 2017).
- We need to find suitable and affordable real estate, e.g., in terms of accessibility, availability of parking spots, desired frequency of passers-by, having big windows, etc.

13.9 Exercises

1. We have used `raster::rasterFromXYZ()` to convert a `input_tidy` into a raster brick. Try to achieve the same with the help of the `sp::gridded()` function.

2. Download the csv file containing inhabitant information for a 100-m cell resolution (`https://www.zensus2011.de/SharedDocs/Downloads/ DE/Pressemitteilung/DemografischeGrunddaten/csv_Bevoelkerung_100m_ Gitter.zip?__blob=publicationFile&v=3`). Please note that the unzipped file has a size of 1.23 GB. To read it into R, you can use `readr::read_csv`. This takes 30 seconds on my machine (16 GB RAM) `data.table::fread()` might be even faster, and returns an object of class `data.table()`. Use `as.tibble()` to convert it into a tibble. Build an inhabitant raster, aggregate it to a cell resolution of 1 km, and compare the difference with the inhabitant raster (`inh`) we have created using class mean values.

3. Suppose our bike shop predominantly sold electric bikes to older people. Change the age raster accordingly, repeat the remaining analyses and compare the changes with our original result.

14

Ecology

Prerequisites

This chapter assumes you have a strong grasp of geographic data analysis and processing, covered in Chapters 2 to 5. In it you will also make use of R's interfaces to dedicated GIS software, and spatial cross-validation, topics covered in Chapters 9 and 11, respectively.

The chapter uses the following packages:

```
library(sf)
library(raster)
library(RQGIS)
library(mlr)
library(dplyr)
library(vegan)
```

14.1 Introduction

In this chapter we will model the floristic gradient of fog oases to reveal distinctive vegetation belts that are clearly controlled by water availability. To do so, we will bring together concepts presented in previous chapters and even extend them (Chapters 2 to 5 and Chapters 9 and 11).

Fog oases are one of the most fascinating vegetation formations we have ever encountered. These formations, locally termed *lomas*, develop on mountains along the coastal deserts of Peru and Chile.[1] The deserts' extreme conditions and remoteness provide the habitat for a unique ecosystem, including species endemic to the fog oases. Despite the arid conditions and low levels of precipitation of around 30-50 mm per year on average, fog deposition increases

---

[1]Similar vegetation formations develop also in other parts of the world, e.g., in Namibia and along the coasts of Yemen and Oman (Galletti et al., 2016).

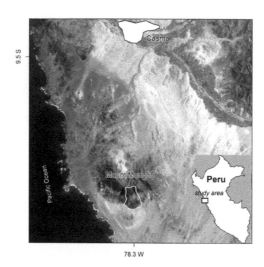

FIGURE 14.1: The Mt. Mongón study area, from Muenchow, Schratz, and Brenning (2017).

the amount of water available to plants during austal winter. This results in green southern-facing mountain slopes along the coastal strip of Peru (Figure 14.1). This fog, which develops below the temperature inversion caused by the cold Humboldt current in austral winter, provides the name for this habitat. Every few years, the El Niño phenomenon brings torrential rainfall to this sun-baked environment (Dillon et al., 2003). This causes the desert to bloom, and provides tree seedlings a chance to develop roots long enough to survive the following arid conditions.

Unfortunately, fog oases are heavily endangered. This is mostly due to human activity (agriculture and climate change). To effectively protect the last remnants of this unique vegetation ecosystem, evidence is needed on the composition and spatial distribution of the native flora (Muenchow et al., 2013a,b). *Lomas* mountains also have economic value as a tourist destination, and can contribute to the well-being of local people via recreation. For example, most Peruvians live in the coastal desert, and *lomas* mountains are frequently the closest "green" destination.

In this chapter we will demonstrate ecological applications of some of the techniques learned in the previous chapters. This case study will involve analyzing the composition and the spatial distribution of the vascular plants on the southern slope of Mt. Mongón, a *lomas* mountain near Casma on the central northern coast of Peru (Figure 14.1).

During a field study to Mt. Mongón, we recorded all vascular plants living in 100 randomly sampled 4x4 m² plots in the austral winter of 2011 (Muenchow et al., 2013a). The sampling coincided with a strong La Niña event that year (see ENSO monitoring of the NOASS Climate Prediction Center[2]). This led to even higher levels of aridity than usual in the coastal desert. On the other hand, it also increased fog activity on the southern slopes of Peruvian *lomas* mountains.

Ordinations are dimension-reducing techniques which allow the extraction of the main gradients from a (noisy) dataset, in our case the floristic gradient developing along the southern mountain slope (see next section). In this chapter we will model the first ordination axis, i.e., the floristic gradient, as a function of environmental predictors such as altitude, slope, catchment area and NDVI. For this, we will make use of a random forest model - a very popular machine learning algorithm (Breiman, 2001). The model will allow us to make spatial predictions of the floristic composition anywhere in the study area. To guarantee an optimal prediction, it is advisable to tune beforehand the hyperparameters with the help of spatial cross-validation (see Section 11.5.2).

14.2   Data and data preparation

All the data needed for the subsequent analyses is available via the **RQGIS** package.

```
data("study_area", "random_points", "comm", "dem", "ndvi", package = "RQGIS")
```

study_area is an sf polygon representing the outlines of the study area. random_points is an sf object, and contains the 100 randomly chosen sites. comm is a community matrix of the wide data format (Wickham, 2014b) where the rows represent the visited sites in the field and the columns the observed species.[3]

```
# sites 35 to 40 and corresponding occurrences of the first five species in the
# community matrix
comm[35:40, 1:5]
#>    Alon_meri Alst_line Alte_hali Alte_porr Anth_eccr
#> 35         0         0         0       0.0     1.000
#> 36         0         0         1       0.0     0.500
#> 37         0         0         0       0.0     0.125
```

---

[2]http://origin.cpc.ncep.noaa.gov/products/analysis_monitoring/ensostuff/ONI_v5.php
[3]In statistics, this is also called a contingency table or cross-table.

```
#> 38        0        0        0        0.0      3.000
#> 39        0        0        0        0.0      2.000
#> 40        0        0        0        0.2      0.125
```

The values represent species cover per site, and were recorded as the area covered by a species in proportion to the site area in percentage points (%; please note that one site can have >100% due to overlapping cover between individual plants). The rownames of comm correspond to the id column of random_points. dem is the digital elevation model (DEM) for the study area, and ndvi is the Normalized Difference Vegetation Index (NDVI) computed from the red and near-infrared channels of a Landsat scene (see Section 4.3.3 and ?ndvi). Visualizing the data helps to get more familiar with it, as shown in Figure 14.2 where the dem is overplotted by the random_points and the study_area.

The next step is to compute variables which we will not only need for the modeling and predictive mapping (see Section 14.4.2) but also for aligning the Non-metric multidimensional scaling (NMDS) axes with the main gradient in the study area, altitude and humidity, respectively (see Section 14.3).

Specifically, we will compute catchment slope and catchment area from a digital elevation model using R-GIS bridges (see Chapter 9). Curvatures might also represent valuable predictors, in the Exercise section you can find out how they would change the modeling result.

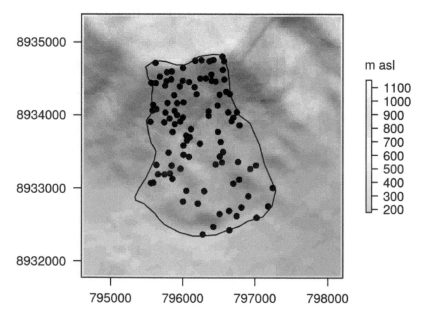

FIGURE 14.2: Study mask (polygon), location of the sampling sites (black points) and DEM in the background.

To compute catchment area and catchment slope, we will make use of the `saga:sagawetnessindex` function.[4] `get_usage()` returns all function parameters and default values of a specific geoalgorithm. Here, we present only a selection of the complete output.

```
get_usage("saga:sagawetnessindex")
#>ALGORITHM: Saga wetness index
#>   DEM <ParameterRaster>
#>   ...
#>   SLOPE_TYPE <ParameterSelection>
#>   ...
#>   AREA <OutputRaster>
#>   SLOPE <OutputRaster>
#>   AREA_MOD <OutputRaster>
#>   TWI <OutputRaster>
#>   ...
#>SLOPE_TYPE(Type of Slope)
#>   0 - [0] local slope
#>   1 - [1] catchment slope
#>   ...
```

Subsequently, we can specify the needed parameters using R named arguments (see Section 9.2). Remember that we can use a `RasterLayer` living in R's global environment to specify the input raster `DEM` (see Section 9.2). Specifying 1 as the `SLOPE_TYPE` makes sure that the algorithm will return the catchment slope. The resulting output rasters should be saved to temporary files with an `.sdat` extension which is a SAGA raster format. Setting `load_output` to `TRUE` ensures that the resulting rasters will be imported into R.

```
# environmental predictors: catchment slope and catchment area
ep = run_qgis(alg = "saga:sagawetnessindex",
              DEM = dem,
              SLOPE_TYPE = 1,
              SLOPE = tempfile(fileext = ".sdat"),
              AREA = tempfile(fileext = ".sdat"),
              load_output = TRUE,
              show_output_paths = FALSE)
```

This returns a list named `ep` consisting of two elements: `AREA` and `SLOPE`. Let us add two more raster objects to the list, namely `dem` and `ndvi`, and convert it into a raster stack (see Section 2.3.3).

---

[4] Admittedly, it is a bit unsatisfying that the only way of knowing that `sagawetnessindex` computes the desired terrain attributes is to be familiar with SAGA.

```
ep = stack(c(dem, ndvi, ep))
names(ep) = c("dem", "ndvi", "carea", "cslope")
```

Additionally, the catchment area values are highly skewed to the right (hist(ep$carea)). A log10-transformation makes the distribution more normal.

```
ep$carea = log10(ep$carea)
```

As a convenience to the reader, we have added ep to **spDataLarge**:

```
data("ep", package = "spDataLarge")
```

Finally, we can extract the terrain attributes to our field observations (see also Section 5.4.2).

```
random_points[, names(ep)] = raster::extract(ep, random_points)
```

14.3   Reducing dimensionality

Ordinations are a popular tool in vegetation science to extract the main information, frequently corresponding to ecological gradients, from large species-plot matrices mostly filled with 0s. However, they are also used in remote sensing, the soil sciences, geomarketing and many other fields. If you are unfamiliar with ordination techniques or in need of a refresher, have a look at Michael W. Palmer's web page[5] for a short introduction to popular ordination techniques in ecology and at Borcard et al. (2011) for a deeper look on how to apply these techniques in R. **vegan**'s package documentation is also a very helpful resource (vignette(package = "vegan")).

Principal component analysis (PCA) is probably the most famous ordination technique. It is a great tool to reduce dimensionality if one can expect linear relationships between variables, and if the joint absence of a variable (for example calcium) in two plots (observations) can be considered a similarity. This is barely the case with vegetation data.

For one, relationships are usually non-linear along environmental gradients. That means the presence of a plant usually follows a unimodal relationship along a gradient (e.g., humidity, temperature or salinity) with a peak at

---

[5]http://ordination.okstate.edu/overview.htm

the most favorable conditions and declining ends towards the unfavorable conditions.

Secondly, the joint absence of a species in two plots is hardly an indication for similarity. Suppose a plant species is absent from the driest (e.g., an extreme desert) and the most moistest locations (e.g., a tree savanna) of our sampling. Then we really should refrain from counting this as a similarity because it is very likely that the only thing these two completely different environmental settings have in common in terms of floristic composition is the shared absence of species (except for rare ubiquitous species).

Non-metric multidimensional scaling (NMDS) is one popular dimension-reducing technique in ecology (von Wehrden et al., 2009). NMDS reduces the rank-based differences between the distances between objects in the original matrix and distances between the ordinated objects. The difference is expressed as stress. The lower the stress value, the better the ordination, i.e., the low-dimensional representation of the original matrix. Stress values lower than 10 represent an excellent fit, stress values of around 15 are still good, and values greater than 20 represent a poor fit (McCune et al., 2002). In R, metaMDS() of the **vegan** package can execute a NMDS. As input, it expects a community matrix with the sites as rows and the species as columns. Often ordinations using presence-absence data yield better results (in terms of explained variance) though the prize is, of course, a less informative input matrix (see also Exercises). decostand() converts numerical observations into presences and absences with 1 indicating the occurrence of a species and 0 the absence of a species. Ordination techniques such as NMDS require at least one observation per site. Hence, we need to dismiss all sites in which no species were found.

```
# presence-absence matrix
pa = decostand(comm, "pa") # 100 rows (sites), 69 columns (species)
# keep only sites in which at least one species was found
pa = pa[rowSums(pa) != 0, ] # 84 rows, 69 columns
```

The resulting output matrix serves as input for the NMDS. k specifies the number of output axes, here, set to 4.[6] NMDS is an iterative procedure trying to make the ordinated space more similar to the input matrix in each step. To make sure that the algorithm converges, we set the number of steps to 500 (try parameter).

```
set.seed(25072018)
nmds = metaMDS(comm = pa, k = 4, try = 500)
nmds$stress
```

---

[6]One way of choosing k is to try k values between 1 and 6 and then using the result which yields the best stress value (McCune et al., 2002).

```
#> ...
#> Run 498 stress 0.08834745
#> ... Procrustes: rmse 0.004100446  max resid 0.03041186
#> Run 499 stress 0.08874805
#> ... Procrustes: rmse 0.01822361  max resid 0.08054538
#> Run 500 stress 0.08863627
#> ... Procrustes: rmse 0.01421176  max resid 0.04985418
#> *** Solution reached
#> 0.08831395
```

A stress value of 9 represents a very good result, which means that the reduced ordination space represents the large majority of the variance of the input matrix. Overall, NMDS puts objects that are more similar (in terms of species composition) closer together in ordination space. However, as opposed to most other ordination techniques, the axes are arbitrary and not necessarily ordered by importance (Borcard et al., 2011). However, we already know that humidity represents the main gradient in the study area (Muenchow et al., 2013a, 2017). Since humidity is highly correlated with elevation, we rotate the NMDS axes in accordance with elevation (see also ?MDSrotate for more details on rotating NMDS axes). Plotting the result reveals that the first axis is, as intended, clearly associated with altitude (Figure 14.3).

```
elev = dplyr::filter(random_points, id %in% rownames(pa)) %>%
  dplyr::pull(dem)
# rotating NMDS in accordance with altitude (proxy for humidity)
rotnmds = MDSrotate(nmds, elev)
# extracting the first two axes
sc = scores(rotnmds, choices = 1:2)
# plotting the first axis against altitude
plot(y = sc[, 1], x = elev, xlab = "elevation in m",
     ylab = "First NMDS axis", cex.lab = 0.8, cex.axis = 0.8)
```

The scores of the first NMDS axis represent the different vegetation formations, i.e., the floristic gradient, appearing along the slope of Mt. Mongón. To spatially visualize them, we can model the NMDS scores with the previously created predictors (Section 14.2), and use the resulting model for predictive mapping (see next section).

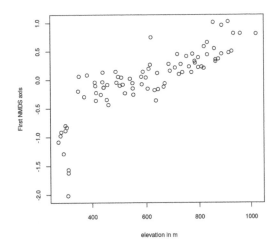

FIGURE 14.3: Plotting the first NMDS axis against altitude.

14.4 Modeling the floristic gradient

To predict the floristic gradient spatially, we will make use of a random forest model (Hengl et al., 2018). Random forest models are frequently used in environmental and ecological modeling, and often provide the best results in terms of predictive performance (Schratz et al., 2018). Here, we shortly introduce decision trees and bagging, since they form the basis of random forests. We refer the reader to James et al. (2013) for a more detailed description of random forests and related techniques.

To introduce decision trees by example, we first construct a response-predictor matrix by joining the rotated NMDS scores to the field observations (random_points). We will also use the resulting data frame for the **mlr** modeling later on.

```
# construct response-predictor matrix
# id- and response variable
rp = data.frame(id = as.numeric(rownames(sc)), sc = sc[, 1])
# join the predictors (dem, ndvi and terrain attributes)
rp = inner_join(random_points, rp, by = "id")
```

Decision trees split the predictor space into a number of regions. To illustrate this, we apply a decision tree to our data using the scores of the first NMDS axis as the response (sc) and altitude (dem) as the only predictor.

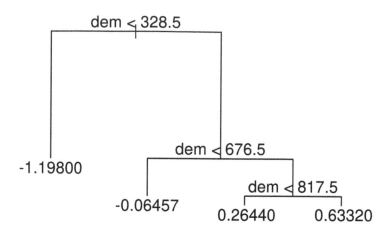

FIGURE 14.4: Simple example of a decision tree with three internal nodes and four terminal nodes.

```
library("tree")
tree_mo = tree(sc ~ dem, data = rp)
plot(tree_mo)
text(tree_mo, pretty = 0)
```

The resulting tree consists of three internal nodes and four terminal nodes (Figure 14.4). The first internal node at the top of the tree assigns all observations which are below 328.5 m to the left and all other observations to the right branch. The observations falling into the left branch have a mean NMDS score of -1.198. Overall, we can interpret the tree as follows: the higher the elevation, the higher the NMDS score becomes. Decision trees have a tendency to overfit, that is they mirror too closely the input data including its noise which in turn leads to bad predictive performances (Section 11.4; James et al., 2013). Bootstrap aggregation (bagging) is an ensemble technique and helps to overcome this problem. Ensemble techniques simply combine the predictions of multiple models. Thus, bagging takes repeated samples from the same input data and averages the predictions. This reduces the variance and overfitting with the result of a much better predictive accuracy compared to decision trees. Finally, random forests extend and improve bagging by decorrelating trees which is desirable since averaging the predictions of highly correlated trees shows a higher variance and thus lower reliability than averaging predictions of decorrelated trees (James et al., 2013). To achieve this, random forests use bagging, but in contrast to the traditional bagging where each tree is allowed to use all available predictors, random forests only use a random sample of all available predictors.

14.4.1 mlr building blocks

The code in this section largely follows the steps we have introduced in Section 11.5.2. The only differences are the following:

1. The response variable is numeric, hence a regression task will replace the classification task of Section 11.5.2.
2. Instead of the AUROC which can only be used for categorical response variables, we will use the root mean squared error (RMSE) as performance measure.
3. We use a random forest model instead of a support vector machine which naturally goes along with different hyperparameters.
4. We are leaving the assessment of a bias-reduced performance measure as an exercise to the reader (see Exercises). Instead we show how to tune hyperparameters for (spatial) predictions.

Remember that 125,500 models were necessary to retrieve bias-reduced performance estimates when using 100-repeated 5-fold spatial cross-validation and a random search of 50 iterations (see Section 11.5.2). In the hyperparameter tuning level, we found the best hyperparameter combination which in turn was used in the outer performance level for predicting the test data of a specific spatial partition (see also Figure 11.6). This was done for five spatial partitions, and repeated a 100 times yielding in total 500 optimal hyperparameter combinations. Which one should we use for making spatial predictions? The answer is simple, none at all. Remember, the tuning was done to retrieve a bias-reduced performance estimate, not to do the best possible spatial prediction. For the latter, one estimates the best hyperparameter combination from the complete dataset. This means, the inner hyperparameter tuning level is no longer needed which makes perfect sense since we are applying our model to new data (unvisited field observations) for which the true outcomes are unavailable, hence testing is impossible in any case. Therefore, we tune the hyperparameters for a good spatial prediction on the complete dataset via a 5-fold spatial CV with one repetition.

The preparation for the modeling using the **mlr** package includes the construction of a response-predictor matrix containing only variables which should be used in the modeling and the construction of a separate coordinate data frame.

```
# extract the coordinates into a separate data frame
coords = sf::st_coordinates(rp) %>%
  as.data.frame() %>%
  rename(x = X, y = Y)
# only keep response and predictors which should be used for the modeling
rp = dplyr::select(rp, -id, -spri) %>%
  st_drop_geometry()
```

Having constructed the input variables, we are all set for specifying the **mlr** building blocks (task, learner, and resampling). We will use a regression task since the response variable is numeric. The learner is a random forest model implementation from the **ranger** package.

```
# create task
task = makeRegrTask(data = rp, target = "sc", coordinates = coords)
# learner
lrn_rf = makeLearner(cl = "regr.ranger", predict.type = "response")
```

As opposed to for example support vector machines (see Section 11.5.2), random forests often already show good performances when used with the default values of their hyperparameters (which may be one reason for their popularity). Still, tuning often moderately improves model results, and thus is worth the effort (Probst et al., 2018). Since we deal with geographic data, we will again make use of spatial cross-validation to tune the hyperparameters (see Sections 11.4 and 11.5). Specifically, we will use a five-fold spatial partitioning with only one repetition (makeResampleDesc()). In each of these spatial partitions, we run 50 models (makeTuneControlRandom()) to find the optimal hyperparameter combination.

```
# spatial partitioning
perf_level = makeResampleDesc("SpCV", iters = 5)
# specifying random search
ctrl = makeTuneControlRandom(maxit = 50L)
```

In random forests, the hyperparameters mtry, min.node.size and sample.fraction determine the degree of randomness, and should be tuned (Probst et al., 2018). mtry indicates how many predictor variables should be used in each tree. If all predictors are used, then this corresponds in fact to bagging (see beginning of Section 14.4). The sample.fraction parameter specifies the fraction of observations to be used in each tree. Smaller fractions lead to greater diversity, and thus less correlated trees which often is desirable (see above). The min.node.size parameter indicates the number of observations a terminal node should at least have (see also Figure 14.4). Naturally, as trees and computing time become larger, the lower the min.node.size.

Hyperparameter combinations will be selected randomly but should fall inside specific tuning limits (makeParamSet()). mtry should range between 1 and the number of predictors (4), sample.fraction should range between 0.2 and 0.9 and min.node.size should range between 1 and 10.

```
# specifying the search space
ps = makeParamSet(
```

```
makeIntegerParam("mtry", lower = 1, upper = ncol(rp) - 1),
makeNumericParam("sample.fraction", lower = 0.2, upper = 0.9),
makeIntegerParam("min.node.size", lower = 1, upper = 10)
)
```

Finally, `tuneParams()` runs the hyperparameter tuning, and will find the optimal hyperparameter combination for the specified parameters. The performance measure is the root mean squared error (RMSE).

```
# hyperparamter tuning
set.seed(02082018)
tune = tuneParams(learner = lrn_rf,
                  task = task,
                  resampling = perf_level,
                  par.set = ps,
                  control = ctrl,
                  measures = mlr::rmse)
#>...
#> [Tune-x] 49: mtry=3; sample.fraction=0.533; min.node.size=5
#> [Tune-y] 49: rmse.test.rmse=0.5636692; time: 0.0 min
#> [Tune-x] 50: mtry=1; sample.fraction=0.68; min.node.size=5
#> [Tune-y] 50: rmse.test.rmse=0.6314249; time: 0.0 min
#> [Tune] Result: mtry=4; sample.fraction=0.887; min.node.size=10 :
#> rmse.test.rmse=0.5104918
```

An `mtry` of 4, a `sample.fraction` of 0.887, and a `min.node.size` of 10 represent the best hyperparameter combination. A RMSE of 0.51 is relatively good when considering the range of the response variable which is 3.04 (`diff(range(rp$sc))`).

14.4.2 Predictive mapping

The tuned hyperparameters can now be used for the prediction. We simply have to modify our learner using the result of the hyperparameter tuning, and run the corresponding model.

```
# learning using the best hyperparameter combination
lrn_rf = makeLearner(cl = "regr.ranger",
                     predict.type = "response",
                     mtry = tune$x$mtry,
                     sample.fraction = tune$x$sample.fraction,
                     min.node.size = tune$x$min.node.size)
# doing the same more elegantly using setHyperPars()
# lrn_rf = setHyperPars(makeLearner("regr.ranger", predict.type = "response"),
```

```
#                        par.vals = tune$x)
# train model
model_rf = train(lrn_rf, task)
# to retrieve the ranger output, run:
# mlr::getLearnerModel(model_rf)
# which corresponds to:
# ranger(sc ~ ., data = rp,
#        mtry = tune$x$mtry,
#        sample.fraction = tune$x$sample.fraction,
#        min.node.sie = tune$x$min.node.size)
```

The last step is to apply the model to the spatially available predictors, i.e., to the raster stack. So far, `raster::predict()` does not support the output of **ranger** models, hence, we will have to program the prediction ourselves. First, we convert `ep` into a prediction data frame which secondly serves as input for the `predict.ranger()` function. Thirdly, we put the predicted values back into a `RasterLayer` (see Section 3.3.1 and Figure 14.5).

```
# convert raster stack into a data frame
new_data = as.data.frame(as.matrix(ep))
# apply the model to the data frame
pred_rf = predict(model_rf, newdata = new_data)
# put the predicted values into a raster
pred = dem
# replace altitudinal values by rf-prediction values
pred[] = pred_rf$data$response
```

The predictive mapping clearly reveals distinct vegetation belts (Figure 14.5). Please refer to Muenchow et al. (2013b) for a detailed description of vegetation belts on **lomas** mountains. The blue color tones represent the so-called *Tillandsia*-belt. *Tillandsia* is a highly adapted genus especially found in high quantities at the sandy and quite desertic foot of *lomas* mountains. The yellow color tones refer to a herbaceous vegetation belt with a much higher plant cover compared to the *Tillandsia*-belt. The orange colors represent the bromeliad belt, which features the highest species richness and plant cover. It can be found directly beneath the temperature inversion (ca. 750-850 m asl) where humidity due to fog is highest. Water availability naturally decreases above the temperature inversion, and the landscape becomes desertic again with only a few succulent species (succulent belt; red colors). Interestingly, the spatial prediction clearly reveals that the bromeliad belt is interrupted which is a very interesting finding we would have not detected without the predictive mapping.

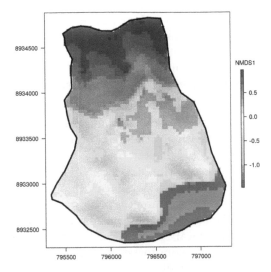

FIGURE 14.5: Predictive mapping of the floristic gradient clearly revealing distinct vegetation belts.

14.5 Conclusions

In this chapter we have ordinated the community matrix of the **lomas** Mt. Mongón with the help of a NMDS (Section 14.3). The first axis, representing the main floristic gradient in the study area, was modeled as a function of environmental predictors which partly were derived through R-GIS bridges (Section 14.2). The **mlr** package provided the building blocks to spatially tune the hyperparameters mtry, sample.fraction and min.node.size (Section 14.4.1). The tuned hyperparameters served as input for the final model which in turn was applied to the environmental predictors for a spatial representation of the floristic gradient (Section 14.4.2). The result demonstrates spatially the astounding biodiversity in the middle of the desert. Since **lomas** mountains are heavily endangered, the prediction map can serve as basis for informed decision-making on delineating protection zones, and making the local population aware of the uniqueness found in their immediate neighborhood.

In terms of methodology, a few additional points could be addressed:

- It would be interesting to also model the second ordination axis, and to subsequently find an innovative way of visualizing jointly the modeled scores of the two axes in one prediction map.
- If we were interested in interpreting the model in an ecologically meaningful

way, we should probably use (semi-)parametric models (Muenchow et al., 2013a; Zuur et al., 2009, 2017). However, there are at least approaches that help to interpret machine learning models such as random forests (see, e.g., https://mlr-org.github.io/interpretable-machine-learning-iml-and-mlr/).

*   A sequential model-based optimization (SMBO) might be preferable to the random search for hyperparameter optimization used in this chapter (Probst et al., 2018).

Finally, please note that random forest and other machine learning models are frequently used in a setting with lots of observations and many predictors, much more than used in this chapter, and where it is unclear which variables and variable interactions contribute to explaining the response. Additionally, the relationships might be highly non-linear. In our use case, the relationship between response and predictors are pretty clear, there is only a slight amount of non-linearity and the number of observations and predictors is low. Hence, it might be worth trying a linear model. A linear model is much easier to explain and understand than a random forest model, and therefore to be preferred (law of parsimony), additionally it is computationally less demanding (see Exercises). If the linear model cannot cope with the degree of non-linearity present in the data, one could also try a generalized additive model (GAM). The point here is that the toolbox of a data scientist consists of more than one tool, and it is your responsibility to select the tool best suited for the task or purpose at hand. Here, we wanted to introduce the reader to random forest modeling and how to use the corresponding results for spatial predictions. For this purpose, a well-studied dataset with known relationships between response and predictors, is appropriate. However, this does not imply that the random forest model has returned the best result in terms of predictive performance (see Exercises).

14.6   Exercises

1. Run a NMDS using the percentage data of the community matrix. Report the stress value and compare it to the stress value as retrieved from the NMDS using presence-absence data. What might explain the observed difference?

2. Compute all the predictor rasters we have used in the chapter (catchment slope, catchment area), and put them into a raster stack. Add dem and ndvi to the raster stack. Next, compute profile and tangential curvature as additional predictor rasters and add them to the raster stack (hint: grass7:r.slope.aspect). Finally, construct a response-predictor matrix. The scores of the first NMDS axis (which were the result when using the presence-absence community

matrix) rotated in accordance with elevation represent the response variable, and should be joined to `random_points` (use an inner join). To complete the response-predictor matrix, extract the values of the environmental predictor raster stack to `random_points`.

3. Use the response-predictor matrix of the previous exercise to fit a random forest model. Find the optimal hyperparameters and use them for making a prediction map.

4. Retrieve the bias-reduced RMSE of a random forest model using spatial cross-validation including the estimation of optimal hyperparameter combinations (random search with 50 iterations) in an inner tuning loop (see Section 11.5.2). Parallelize the tuning level (see Section 11.5.2). Report the mean RMSE and use a boxplot to visualize all retrieved RMSEs.

5. Retrieve the bias-reduced RMSE of a simple linear model using spatial cross-validation. Compare the result to the result of the random forest model by making RMSE boxplots for each modeling approach.

15

Conclusion

Prerequisites

Like the introduction, this concluding chapter contains few code chunks. But its prerequisites are demanding. It assumes that you have:

- Read through and attempted the exercises in all the chapters of Part I (Foundations).
- Grasped the diversity of methods that build on these foundations, by following the code and prose in Part II (Extensions).
- Considered how you can use geocomputation to solve real-world problems, at work and beyond, after engaging with Part III (Applications).

15.1 Introduction

The aim of this chapter is to synthesize the contents, with reference to recurring themes/concepts, and to inspire future directions of application and development. Section 15.2 discusses the wide range of options for handling geographic data in R. Choice is a key feature of open source software; the section provides guidance on choosing between the various options. Section 15.3 describes gaps in the book's contents and explains why some areas of research were deliberately omitted, while others were emphasized. This discussion leads to the question (which is answered in Section 15.4): having read this book, where to go next? Section 15.5 returns to the wider issues raised in Chapter 1. In it we consider geocomputation as part of a wider 'open source approach' that ensures methods are publicly accessible, reproducible and supported by collaborative communities. This final section of the book also provides some pointers on how to get involved.

15.2 Package choice

A characteristic of R is that there are often multiple ways to achieve the same result. The code chunk below illustrates this by using three functions, covered in Chapters 3 and 5, to combine the 16 regions of New Zealand into a single geometry:

```
library(spData)
nz_u1 = sf::st_union(nz)
nz_u2 = aggregate(nz["Population"], list(rep(1, nrow(nz))), sum)
nz_u3 = dplyr::summarise(nz, t = sum(Population))
identical(nz_u1, nz_u2$geometry)
#> [1] TRUE
identical(nz_u1, nz_u3$geom)
#> [1] TRUE
```

Although the classes, attributes and column names of the resulting objects nz_u1 to nz_u3 differ, their geometries are identical. This is verified using the base R function identical().[1] Which to use? It depends: the former only processes the geometry data contained in nz so is faster, while the other options performed attribute operations, which may be useful for subsequent steps.

The wider point is that there are often multiple options to choose from when working with geographic data in R, even within a single package. The range of options grows further when more R packages are considered: you could achieve the same result using the older **sp** package, for example. We recommend using **sf** and the other packages showcased in this book, for reasons outlined in Chapter 2, but it's worth being aware of alternatives and being able to justify your choice of software.

A common (and sometimes controversial) choice is between **tidyverse** and base R approaches. We cover both and encourage you to try both before deciding which is more appropriate for different tasks. The following code chunk, described in Chapter 3, shows how attribute data subsetting works in each approach, using the base R operator [and the select() function from the **tidyverse** package **dplyr**. The syntax differs but the results are (in essence) the same:

---

[1]The first operation, undertaken by the function st_union(), creates an object of class sfc (a simple feature column). The latter two operations create sf objects, each of which *contains* a simple feature column. Therefore, it is the geometries contained in simple feature columns, not the objects themselves, that are identical.

```
library(dplyr)                          # attach tidyverse package
nz_name1 = nz["Name"]                   # base R approach
nz_name2 = nz %>% select(Name)          # tidyverse approach
identical(nz_name1$Name, nz_name2$Name) # check results
#> [1] TRUE
```

Again the question arises: which to use? Again the answer is: it depends. Each approach has advantages: the pipe syntax is popular and appealing to some, while base R is more stable, and is well known to others. Choosing between them is therefore largely a matter of preference. However, if you do choose to use **tidyverse** functions to handle geographic data, beware of a number of pitfalls (see the supplementary article tidyverse-pitfalls[2] on the website that supports this book).

While commonly needed operators/functions were covered in depth — such as the base R [subsetting operator and the **dplyr** function filter() — there are many other functions for working with geographic data, from other packages, that have not been mentioned. Chapter 1 mentions 20+ influential packages for working with geographic data, and only a handful of these are demonstrated in subsequent chapters. There are hundreds more. As of early 2019, there are nearly 200 packages mentioned in the Spatial Task View[3]; more packages and countless functions for geographic data are developed each year, making it impractical to cover them all in a single book.

The rate of evolution in R's spatial ecosystem may seem overwhelming, but there are strategies to deal with the wide range of options. Our advice is to start by learning one approach *in depth* but to have a general understand of the *breadth* of options available. This advice applies equally to solving geographic problems in R (Section 15.4 covers developments in other languages) as it does to other fields of knowledge and application.

Of course, some packages perform much better than others, making package selection an important decision. From this diversity, we have focused on packages that are future-proof (they will work long into the future), high performance (relative to other R packages) and complementary. But there is still overlap in the packages we have used, as illustrated by the diversity of packages for making maps, for example (see Chapter 8).

Package overlap is not necessarily a bad thing. It can increase resilience, performance (partly driven by friendly competition and mutual learning between developers) and choice, a key feature of open source software. In this context the decision to use a particular approach, such as the **sf/tidyverse/raster** ecosystem advocated in this book should be made with knowledge of alternatives. The **sp/rgdal/rgeos** ecosystem that **sf** is designed to supersede, for

---

[2]https://geocompr.github.io/geocompkg/articles/tidyverse-pitfalls.html
[3]https://cran.r-project.org/web/views/

example, can do many of the things covered in this book and, due to its age, is built on by many other packages.[4] Although best known for point pattern analysis, the **spatstat** package also supports raster and other vector geometries (Baddeley and Turner, 2005). At the time of writing (October 2018) 69 packages depend on it, making it more than a package: **spatstat** is an alternative R-spatial ecosystem.

It is also being aware of promising alternatives that are under development. The package **stars**, for example, provides a new class system for working with spatiotemporal data. If you are interested in this topic, you can check for updates on the package's source code[5] and the broader SpatioTemporal Task View[6]. The same principle applies to other domains: it is important to justify software choices and review software decisions based on up-to-date information.

15.3   Gaps and overlaps

There are a number of gaps in, and some overlaps between, the topics covered in this book. We have been selective, emphasizing some topics while omitting others. We have tried to emphasize topics that are most commonly needed in real-world applications such as geographic data operations, projections, data read/write and visualization. These topics appear repeatedly in the chapters, a substantial area of overlap designed to consolidate these essential skills for geocomputation.

On the other hand, we have omitted topics that are less commonly used, or which are covered in-depth elsewhere. Statistical topics including point pattern analysis, spatial interpolation (kriging) and spatial epidemiology, for example, are only mentioned with reference to other topics such as the machine learning techniques covered in Chapter 11 (if at all). There is already excellent material on these methods, including statistically orientated chapters in Bivand et al. (2013) and a book on point pattern analysis by Baddeley et al. (2015). Other topics which received limited attention were remote sensing and using R alongside (rather than as a bridge to) dedicated GIS software. There are many resources on these topics, including Wegmann et al. (2016) and the GIS-related teaching materials available from Marburg University[7].

Instead of covering spatial statistical modeling and inference techniques, we focussed on machine learning (see Chapters 11 and 14). Again, the reason

---

[4]At the time of writing 452 package Depend or Import **sp**, showing that its data structures are widely used and have been extended in many directions. The equivalent number for **sf** was 69 in October 2018; with the growing popularity of **sf**, this is set to grow.

[5]https://github.com/r-spatial/stars

[6]https://cran.r-project.org/web/views/SpatioTemporal.html

[7]https://moc.online.uni-marburg.de/doku.php

was that there are already excellent resources on these topics, especially with ecological use cases, including Zuur et al. (2009), Zuur et al. (2017) and freely available teaching material and code on *Geostatistics & Open-source Statistical Computing* by David Rossiter, hosted at css.cornell.edu/faculty/dgr2[8]. There are also excellent resources on spatial statistics using Bayesian modeling, a powerful framework for modeling and uncertainty estimation (Blangiardo and Cameletti, 2015; Krainski et al., 2018).

Finally, we have largely omitted big data analytics. This might seem surprising since especially geographic data can become big really fast. But the prerequisite for doing big data analytics is to know how to solve a problem on a small dataset. Once you have learned that, you can apply the exact same techniques on big data questions, though of course you need to expand your toolbox. The first thing to learn is to handle geographic data queries. This is because big data analytics often boil down to extracting a small amount of data from a database for a specific statistical analysis. For this, we have provided an introduction to spatial databases and how to use a GIS from within R in Chapter 9. If you really have to do the analysis on a big or even the complete dataset, hopefully, the problem you are trying to solve is embarrassingly parallel. For this, you need to learn a system that is able to do this parallelization efficiently such as Hadoop, GeoMesa (http://www.geomesa.org/) or GeoSpark (Huang et al., 2017). But still, you are applying the same techniques and concepts you have used on small datasets to answer a big data question, the only difference is that you then do it in a big data setting.

15.4 Where to go next?

As indicated in the previous sections, the book has covered only a fraction of the R's geographic ecosystem, and there is much more to discover. We have progressed quickly, from geographic data models in Chapter 2, to advanced applications in Chapter 14. Consolidation of skills learned, discovery of new packages and approaches for handling geographic data, and application of the methods to new datasets and domains are suggested future directions. This section expands on this general advice by suggesting specific 'next steps', highlighted in **bold** below.

In addition to learning about further geographic methods and applications with R, for example with reference to the work cited in the previous section, deepening your understanding of **R itself** is a logical next step. R's fundamental classes such as `data.frame` and `matrix` are the foundation of `sf` and `raster` classes, so studying them will improve your understanding of geographic data. This

---

[8]http://www.css.cornell.edu/faculty/dgr2/teach/degeostats.html

can be done with reference to documents that are part of R, and which can be found with the command `help.start()` and additional resources on the subject such as those by Wickham (2014a) and Chambers (2016).

Another software-related direction for future learning is **discovering geocomputation with other languages**. There are good reasons for learning R as a language for geocomputation, as described in Chapter 1, but it is not the only option.[9] It would be possible to study *Geocomputation with: Python, C++, JavaScript, Scala* or *Rust* in equal depth. Each has evolving geospatial capabilities. **rasterio**[10], for example, is a Python package that could supplement/replace the **raster** package used in this book — see Garrard (2016) and online tutorials such as automating-gis-processes[11] for more on the Python ecosystem. Dozens of geospatial libraries have been developed in C++, including well known libraries such as GDAL and GEOS, and less well known libraries such as the **Orfeo Toolbox**[12] for processing remote sensing (raster) data. **Turf.js**[13] is an example of the potential for doing geocomputation with JavaScript. GeoTrellis[14] provides functions for working with raster and vector data in the Java-based language Scala. And WhiteBoxTools[15] provides an example of a rapidly evolving command-line GIS implemented in Rust. Each of these packages/libraries/languages has advantages for geocomputation and there are many more to discover, as documented in the curated list of open source geospatial resources Awesome-Geospatial[16].

There is more to geocomputation than software, however. We can recommend **exploring and learning new research topics and methods** from academic and theoretical perspectives. Many methods that have been written about have yet to be implemented. Learning about geographic methods and potential applications can therefore be rewarding, before writing any code. An example of geographic methods that are increasingly implemented in R is sampling strategies for scientific applications. A next step in this case is to read-up on relevant articles in the area such as Brus (2018), which is accompanied by reproducible code and tutorial content hosted at github.com/DickBrus/TutorialSampling4DSM[17].

---

[9]R's strengths relevant to our definition of geocomputation include its emphasis on scientific reproducibility, widespread use in academic research and unparalleled support for statistical modeling of geographic data. Furthermore, we advocate learning one language (R) for geocomputation in depth before delving into other languages/frameworks because of the costs associated with context switching. It is preferable to have expertise in one language than basic knowledge of many.

[10]https://github.com/mapbox/rasterio

[11]https://automating-gis-processes.github.io/CSC18

[12]https://github.com/orfeotoolbox/OTB

[13]https://github.com/Turfjs/turf

[14]https://geotrellis.io/

[15]https://github.com/jblindsay/whitebox-tools

[16]https://github.com/sacridini/Awesome-Geospatial

[17]https://github.com/DickBrus/TutorialSampling4DSM

15.5 The open source approach

This is a technical book so it makes sense for the next steps, outlined in the previous section, to also be technical. However, there are wider issues worth considering in this final section, which returns to our definition of geocomputation. One of the elements of the term introduced in Chapter 1 was that geographic methods should have a positive impact. Of course, how to define and measure 'positive' is a subjective, philosophical question, beyond the scope of this book. Regardless of your worldview, consideration of the impacts of geocomputational work is a useful exercise: the potential for positive impacts can provide a powerful motivation for future learning and, conversely, new methods can open-up many possible fields of application. These considerations lead to the conclusion that geocomputation is part of a wider 'open source approach'.

Section 1.1 presented other terms that mean roughly the same thing as geocomputation, including geographic data science (GDS) and 'GIScience'. Both capture the essence of working with geographic data, but geocomputation has advantages: it concisely captures the 'computational' way of working with geographic data advocated in this book — implemented in code and therefore encouraging reproducibility — and builds on desirable ingredients of its early definition (Openshaw and Abrahart, 2000):

- The *creative* use of geographic data.
- Application to *real-world problems*.
- Building 'scientific' tools.
- Reproducibility.

We added the final ingredient: reproducibility was barely mentioned in early work on geocomputation, yet a strong case can be made for it being a vital component of the first two ingredients. Reproducibility:

- Encourages *creativity* by encouraging the focus to shift away from the basics (which are readily available through shared code) and towards applications.
- Discourages people from 'reinventing the wheel': there is no need to re-do what others have done if their methods can be used by others.
- Makes academic research more conducive to real world applications, by methods developed for one purpose (perhaps purely academic) can be used for practical applications.

If reproducibility is the defining feature of geocomputation (or command-line GIS, code-driven geographic data analysis, or any other synonym for the same thing) it is worth considering what makes it reproducible. This brings us to the 'open source approach', which has three main components:

- A command-line interface (CLI), encouraging scripts recording geographic work to be shared and reproduced.
- Open source software, which can be inspected and potentially improved by anyone in the world.
- An active developer community, which collaborates and self-organizes to build complementary and modular tools.

Like the term geocomputation, the open source approach is more than a technical entity. It is a community composed of people interacting daily with shared aims: to produce high performance tools, free from commercial or legal restrictions, that are accessible for anyone to use. The open source approach to working with geographic data has advantages that transcend the technicalities of how the software works, encouraging learning, collaboration and an efficient division of labor.

There are many ways to engage in this community, especially with the emergence of code hosting sites, such as GitHub, which encourage communication and collaboration. A good place to start is simply browsing through some of the source code, 'issues' and 'commits' in a geographic package of interest. A quick glance at the r-spatial/sf GitHub repository, which hosts the code underlying the **sf** package, shows that 40+ people have contributed to the codebase and documentation. Dozens more people have contributed by asking question and by contributing to 'upstream' packages that **sf** uses. More than 600 issues have been closed on its issue tracker[18], representing a huge amount of work to make **sf** faster, more stable and user-friendly. This example, from just one package out of dozens, shows the scale of the intellectual operation underway to make R a highly effective and continuously evolving language for geocomputation.

It is instructive to watch the incessant development activity happen in public fora such as GitHub, but it is even more rewarding to become an active participant. This is one of the greatest features of the open source approach: it encourages people to get involved. This book itself is a result of the open source approach: it was motivated by the amazing developments in R's geographic capabilities over the last two decades, but made practically possible by dialogue and code sharing on platforms for collaboration. We hope that in addition to disseminating useful methods for working with geographic data, this book inspires you to take a more open source approach. Whether it's raising a constructive issue alerting developers to problems in their package; making the work done by you and the organizations you work for open; or simply helping other people by passing on the knowledge you've learned, getting involved can be a rewarding experience.

---

[18]https://github.com/r-spatial/sf/issues

Bibliography

Abelson, H., Sussman, G. J., and Sussman, J. (1996). *Structure and Interpretation of Computer Programs*. The MIT Electrical Engineering and Computer Science Series. MIT Press, Cambridge, Massachusetts, second edition.

Akima, H. and Gebhardt, A. (2016). *Akima: Interpolation of Irregularly and Regularly Spaced Data*.

Baddeley, A., Rubak, E., and Turner, R. (2015). *Spatial Point Patterns: Methodology and Applications with R*. CRC Press.

Baddeley, A. and Turner, R. (2005). Spatstat: An R package for analyzing spatial point patterns. *Journal of statistical software*, 12(6):1–42.

Bellos, A. (2011). *Alex's Adventures in Numberland*. Bloomsbury Paperbacks, London.

Bischl, B., Lang, M., Kotthoff, L., Schiffner, J., Richter, J., Studerus, E., Casalicchio, G., and Jones, Z. M. (2016). Mlr: Machine Learning in R. *Journal of Machine Learning Research*, 17(170):1–5.

Bivand, R. (2001). More on Spatial Data Analysis. *R News*, 1(3):13–17.

Bivand, R. (2003). Approaches to Classes for Spatial Data in R. In Hornik, K., Leisch, F., and Zeileis, A., editors, *Proceedings of DSC*.

Bivand, R. (2016a). *Rgrass7: Interface Between GRASS 7 Geographical Information System and R*.

Bivand, R. (2016b). *Spgrass6: Interface between GRASS 6 and R*.

Bivand, R. (2017). *Spdep: Spatial Dependence: Weighting Schemes, Statistics and Models*.

Bivand, R. and Gebhardt, A. (2000). Implementing functions for spatial statistical analysis using the language. *Journal of Geographical Systems*, 2(3):307–317.

Bivand, R., Keitt, T., and Rowlingson, B. (2018). *rgdal: Bindings for the 'Geospatial' Data Abstraction Library*. R package version 1.3-3.

Bivand, R. and Lewin-Koh, N. (2017). *Maptools: Tools for Reading and Handling Spatial Objects*.

Bivand, R. and Neteler, M. (2000). Open source geocomputation: Using the R data analysis language integrated with GRASS GIS and PostgreSQL data base systems. In Neteler, M. and Bivand, R. S., editors, *Proceedings of the 5th International Conference on GeoComputation*.

Bivand, R., Pebesma, E. J., and Gómez-Rubio, V. (2013). *Applied Spatial Data Analysis with R*, volume 747248717. Springer.

Bivand, R. and Rundel, C. (2018). *rgeos: Interface to Geometry Engine - Open Source ('GEOS')*. R package version 0.3-28.

Bivand, R. S. (2000). Using the R statistical data analysis language on GRASS 5.0 GIS database files. *Computers & Geosciences*, 26(9):1043–1052.

Blangiardo, M. and Cameletti, M. (2015). *Spatial and Spatio-Temporal Bayesian Models with R-INLA*. John Wiley & Sons, Ltd, Chichester, UK.

Borcard, D., Gillet, F., and Legendre, P. (2011). *Numerical Ecology with R*. Use R! Springer, New York. OCLC: ocn690089213.

Borland, D. and Taylor II, R. M. (2007). Rainbow color map (still) considered harmful. *IEEE computer graphics and applications*, 27(2).

Breiman, L. (2001). Random Forests. *Machine Learning*, 45(1):5–32.

Brenning, A. (2012a). *ArcGIS Geoprocessing in R via Python*.

Brenning, A. (2012b). Spatial cross-validation and bootstrap for the assessment of prediction rules in remote sensing: The R package sperrorest. pages 5372–5375. IEEE.

Brenning, A., Bangs, D., and Becker, M. (2018). *RSAGA: SAGA Geoprocessing and Terrain Analysis*. R package version 1.1.0.

Brewer, C. A. (2015). *Designing Better Maps: A Guide for GIS Users*. Esri Press, Redlands, California, second edition.

Bristol City Council (2015). Deprivation in Bristol 2015. Technical report, Bristol City Council.

Brus, D. J. (2018). Sampling for digital soil mapping: A tutorial supported by R scripts. *Geoderma*.

Brzustowicz, M. R. (2017). *Data Science with Java: [Practical Methods for Scientists and Engineers]*. O´Reilly, Beijing Boston Farnham, first edition. OCLC: 993428657.

Bucklin, D. and Basille, M. (2018). Rpostgis: Linking R with a PostGIS Spatial Database. *The R Journal*.

Burrough, P. A., McDonnell, R., and Lloyd, C. D. (2015). *Principles of Geographical Information Systems*. Oxford University Press, Oxford, New York, third edition. OCLC: ocn915100245.

Calenge, C. (2006). The package adehabitat for the R software: Tool for the analysis of space and habitat use by animals. *Ecological Modelling*, 197:1035.

Cawley, G. C. and Talbot, N. L. (2010). On over-fitting in model selection and subsequent selection bias in performance evaluation. *Journal of Machine Learning Research*, 11(Jul):2079–2107.

Chambers, J. M. (2016). *Extending R*. CRC Press.

Cheshire, J. and Lovelace, R. (2015). Spatial data visualisation with R. In Brunsdon, C. and Singleton, A., editors, *Geocomputation*, pages 1–14. SAGE Publications.

Conrad, O., Bechtel, B., Bock, M., Dietrich, H., Fischer, E., Gerlitz, L., Wehberg, J., Wichmann, V., and Böhner, J. (2015). System for Automated Geoscientific Analyses (SAGA) v. 2.1.4. *Geosci. Model Dev.*, 8(7):1991–2007.

Coombes, M. G., Green, A. E., and Openshaw, S. (1986). An Efficient Algorithm to Generate Official Statistical Reporting Areas: The Case of the 1984 Travel-to-Work Areas Revision in Britain. *The Journal of the Operational Research Society*, 37(10):943.

Coppock, J. T. and Rhind, D. W. (1991). The history of GIS. *Geographical Information Systems: Principles and Applications, vol. 1.*, 1(1):21–43.

de Berg, M., Cheong, O., van Kreveld, M., and Overmars, M. (2008). *Computational Geometry: Algorithms and Applications*. Springer Science & Business Media.

Diggle, P. and Ribeiro, P. J. (2007). *Model-Based Geostatistics*. Springer.

Dillon, M. O., Nakazawa, M., and Leiva, S. G. (2003). The Lomas formations of coastal Peru: Composition and biogeographic history. In Haas, J. and Dillon, M. O., editors, *El Niño in Peru: Biology and Culture over 10,000 Years*, pages 1–9. Field Museum of Natural History, Chicago.

Douglas, D. H. and Peucker, T. K. (1973). Algorithms for the reduction of the number of points required to represent a digitized line or its caricature. *Cartographica: The International Journal for Geographic Information and Geovisualization*, 10(2):112–122.

Eddelbuettel, D. and Balamuta, J. J. (2018). Extending R with C++: A Brief Introduction to Rcpp. *The American Statistician*, 72(1):28–36.

Galletti, C. S., Turner, B. L., and Myint, S. W. (2016). Land changes and their drivers in the cloud forest and coastal zone of Dhofar, Oman, between 1988 and 2013. *Regional Environmental Change*, 16(7):2141–2153.

Garrard, C. (2016). *Geoprocessing with Python*. Manning Publications, Shelter Island, NY. OCLC: ocn915498655.

Gelfand, A. E., Diggle, P., Guttorp, P., and Fuentes, M. (2010). *Handbook of Spatial Statistics*. CRC press.

Gillespie, C. and Lovelace, R. (2016). *Efficient R Programming: A Practical Guide to Smarter Programming*. O'Reilly Media.

Goovaerts, P. (1997). *Geostatistics for Natural Resources Evaluation*. Applied Geostatistics Series. Oxford University Press, New York.

Graser, A. and Olaya, V. (2015). Processing: A Python Framework for the Seamless Integration of Geoprocessing Tools in QGIS.

Grolemund, G. and Wickham, H. (2016). *R for Data Science*. O'Reilly Media.

Hengl, T. (2007). *A Practical Guide to Geostatistical Mapping of Environmental Variables*. Publications Office, Luxembourg. OCLC: 758643236.

Hengl, T., Nussbaum, M., Wright, M. N., Heuvelink, G. B., and Gräler, B. (2018). Random forest as a generic framework for predictive modeling of spatial and spatio-temporal variables. *PeerJ*, 6:e5518.

Hickman, R., Ashiru, O., and Banister, D. (2011). Transitions to low carbon transport futures: Strategic conversations from London and Delhi. *Journal of Transport Geography*, 19(6):1553–1562.

Hijmans, R. J. (2016). *Geosphere: Spherical Trigonometry*.

Hijmans, R. J. (2017). *raster: Geographic Data Analysis and Modeling*. R package version 2.6-7.

Hollander, Y. (2016). *Transport Modelling for a Complete Beginner*. CTthink!

Horni, A., Nagel, K., and Axhausen, K. W. (2016). *The Multi-Agent Transport Simulation MATSim*. Ubiquity Press.

Huang, Z., Chen, Y., Wan, L., and Peng, X. (2017). GeoSpark SQL: An Effective Framework Enabling Spatial Queries on Spark. *ISPRS International Journal of Geo-Information*, 6(9):285.

Huff, D. L. (1963). A Probabilistic Analysis of Shopping Center Trade Areas. *Land Economics*, 39(1):81–90.

Hunziker, P. (2017). *Velox: Fast Raster Manipulation and Extraction*.

Jafari, E., Gemar, M. D., Juri, N. R., and Duthie, J. (2015). Investigation of Centroid Connector Placement for Advanced Traffic Assignment Models with Added Network Detail. *Transportation Research Record: Journal of the Transportation Research Board*, 2498:19–26.

James, G., Witten, D., Hastie, T., and Tibshirani, R., editors (2013). *An Introduction to Statistical Learning: With Applications in R*. Number 103 in Springer Texts in Statistics. Springer, New York. OCLC: ocn828488009.

Jenny, B., Šavrič, B., Arnold, N. D., Marston, B. E., and Preppernau, C. A. (2017). A guide to selecting map projections for world and hemisphere maps. In Lapaine, M. and Usery, L., editors, *Choosing a Map Projection*, pages 213–228. Springer.

Jr, P. J. R. and Diggle, P. J. (2016). *geoR: Analysis of Geostatistical Data*.

Kahle, D. and Wickham, H. (2013). Ggmap: Spatial Visualization with ggplot2. *The R Journal*, 5:144–161.

Kaiser, M. and Morin, T. (1993). Algorithms for computing centroids. *Computers & Operations Research*, 20(2):151–165.

Karatzoglou, A., Smola, A., Hornik, K., and Zeileis, A. (2004). Kernlab - An S4 Package for Kernel Methods in R. *Journal of Statistical Software*, 11(9).

Knuth, D. E. (1974). Computer Programming As an Art. *Commun. ACM*, 17(12):667–673.

Krainski, E., Gómez Rubio, V., Bakka, H., Lenzi, A., Castro-Camilo, D., Simpson, D., Lindgren, F., and Rue, H. (2018). *Advanced Spatial Modeling with Stochastic Partial Differential Equations Using R and INLA*.

Krug, R. M., Roura-Pascual, N., and Richardson, D. M. (2010). Clearing of invasive alien plants under different budget scenarios: Using a simulation model to test efficiency. *Biological invasions*, 12(12):4099–4112.

Kuhn, M. and Johnson, K. (2013). *Applied Predictive Modeling*. Springer, New York. OCLC: ocn827083441.

Lamigueiro, O. P. (2018). *Displaying Time Series, Spatial, and Space-Time Data with R*. Chapman and Hall/CRC, Boca Raton, second edition.

Landa, M. (2008). New GUI for GRASS GIS based on wxPython. *Departament of Geodesy and Cartography*, pages 1–17.

Liu, J.-G. and Mason, P. J. (2009). *Essential Image Processing and GIS for Remote Sensing*. Wiley-Blackwell, Chichester, West Sussex, UK, Hoboken, NJ.

Livingstone, D. N. (1992). *The Geographical Tradition: Episodes in the History of a Contested Enterprise*. John Wiley & Sons Ltd, Oxford, UK ; Cambridge, USA.

Longley, P. (2015). *Geographic Information Science & Systems*. Wiley, Hoboken, NJ, fourth edition.

Longley, P. A., Brooks, S. M., McDonnell, R., and MacMillan, B., editors (1998). *Geocomputation: A Primer*. Wiley, Chichester, Eng. ; New York.

Lovelace, R. and Dumont, M. (2016). *Spatial Microsimulation with R*. CRC Press.

Lovelace, R., Goodman, A., Aldred, R., Berkoff, N., Abbas, A., and Woodcock, J. (2017). The Propensity to Cycle Tool: An open source online system for sustainable transport planning. *Journal of Transport and Land Use*, 10(1).

Majure, J. J. and Gebhardt, A. (2016). *Sgeostat: An Object-Oriented Framework for Geostatistical Modeling in S+*.

Maling, D. H. (1992). *Coordinate Systems and Map Projections.* Pergamon Press, Oxford ; New York, second edition.

McCune, B., Grace, J. B., and Urban, D. L. (2002). *Analysis of Ecological Communities.* MjM Software Design, Gleneden Beach, OR, second edition. OCLC: 846056595.

Meulemans, W., Dykes, J., Slingsby, A., Turkay, C., and Wood, J. (2017). Small Multiples with Gaps. *IEEE Transactions on Visualization and Computer Graphics*, 23(1):381–390.

Meyer, H., Reudenbach, C., Hengl, T., Katurji, M., and Nauss, T. (2018). Improving performance of spatio-temporal machine learning models using forward feature selection and target-oriented validation. *Environmental Modelling & Software*, 101:1–9.

Miller, H. J. (2004). Tobler's first law and spatial analysis. *Annals of the Association of American Geographers*, 94(2).

Moreno-Monroy, A. I., Lovelace, R., and Ramos, F. R. (2017). Public transport and school location impacts on educational inequalities: Insights from São Paulo. *Journal of Transport Geography.*

Muenchow, J., Bräuning, A., Rodríguez, E. F., and von Wehrden, H. (2013a). Predictive mapping of species richness and plant species' distributions of a Peruvian fog oasis along an altitudinal gradient. *Biotropica*, 45(5):557–566.

Muenchow, J., Brenning, A., and Richter, M. (2012). Geomorphic process rates of landslides along a humidity gradient in the tropical Andes. *Geomorphology*, 139-140:271–284.

Muenchow, J., Dieker, P., Kluge, J., Kessler, M., and von Wehrden, H. (2018). A review of ecological gradient research in the Tropics: Identifying research gaps, future directions, and conservation priorities. *Biodiversity and Conservation*, 27(2):273–285.

Muenchow, J., Hauenstein, S., Bräuning, A., Bäumler, R., Rodríguez, E. F., and von Wehrden, H. (2013b). Soil texture and altitude, respectively, largely determine the floristic gradient of the most diverse fog oasis in the Peruvian desert. *Journal of Tropical Ecology*, 29(05):427–438.

Muenchow, J., Schratz, P., and Brenning, A. (2017). RQGIS: Integrating R with QGIS for statistical geocomputing. *The R Journal*, 9(2):409–428.

Murrell, P. (2016). *R Graphics.* CRC Press, second edition.

Neteler, M. and Mitasova, H. (2008). *Open Source GIS: A GRASS GIS Approach.* Springer, New York, NY, third edition. OCLC: 255568974.

Nolan, D. and Lang, D. T. (2014). *XML and Web Technologies for Data Sciences with R.* Use R! Springer, New York, NY. OCLC: 841520665.

Obe, R. O. and Hsu, L. S. (2015). *PostGIS in Action*. Manning, Shelter Island, NY, second edition. OCLC: ocn872985108.

Office for National Statistics (2014). Workplace Zones: A new geography for workplace statistics - Datasets. https://data.gov.uk/dataset/workplace-zones-a-new-geography-for-workplace-statistics3.

Openshaw, S. and Abrahart, R. J., editors (2000). *Geocomputation*. CRC Press, London ; New York.

O'Rourke, J. (1998). *Computational Geometry in C*. Cambridge University Press, Cambridge, UK, ; New York, NY, USA, second edition.

Padgham, M., Rudis, B., Lovelace, R., and Salmon, M. (2018). *osmdata: Import 'OpenStreetMap' Data as Simple Features or Spatial Objects*. R package version 0.0.7.

Pebesma, E. (2018). Simple features for R: Standardized support for spatial vector data. *The R Journal*.

Pebesma, E. and Bivand, R. (2018). *sp: Classes and Methods for Spatial Data*. R package version 1.3-1.

Pebesma, E. and Graeler, B. (2018). *gstat: Spatial and Spatio-Temporal Geostatistical Modelling, Prediction and Simulation*. R package version 1.1-6.

Pebesma, E., Mailund, T., and Hiebert, J. (2016). Measurement Units in R. *The R Journal*, 8(2):486–494.

Pebesma, E., Nüst, D., and Bivand, R. (2012). The R software environment in reproducible geoscientific research. *Eos, Transactions American Geophysical Union*, 93(16):163–163.

Pebesma, E. J. and Bivand, R. S. (2005). Classes and methods for spatial data in R. *R news*, 5(2):9–13.

Pezanowski, S., MacEachren, A. M., Savelyev, A., and Robinson, A. C. (2018). SensePlace3: A geovisual framework to analyze place–time–attribute information in social media. *Cartography and Geographic Information Science*, 45(5):420–437.

Probst, P., Wright, M., and Boulesteix, A.-L. (2018). Hyperparameters and Tuning Strategies for Random Forest. *arXiv:1804.03515 [cs, stat]*.

Qiu, F., Zhang, C., and Zhou, Y. (2012). The Development of an Areal Interpolation ArcGIS Extension and a Comparative Study. *GIScience & Remote Sensing*, 49(5):644–663.

Ripley, B. D. (2001). Spatial Statistics in R. *R News*, 1(2):14–15.

Rodrigue, J.-P., Comtois, C., and Slack, B. (2013). *The Geography of Transport Systems*. Routledge, London, New York, third edition.

Rowlingson, B., Baddeley, A., Turner, R., and Diggle, P. (2003). Rasp: A

Package for Spatial Statistics. In Hornik, K., editor, *Proceedings of the 3rd International Workshop on Distributed Statistical Computing*.

Rowlingson, B. and Diggle, P. (2017). *Splancs: Spatial and Space-Time Point Pattern Analysis*.

Rowlingson, B. S. and Diggle, P. J. (1993). Splancs: Spatial point pattern analysis code in S-plus. *Computers & Geosciences*, 19(5):627–655.

Schratz, P., Muenchow, J., Iturritxa, E., Richter, J., and Brenning, A. (2018). Performance evaluation and hyperparameter tuning of statistical and machine-learning models using spatial data. *arXiv:1803.11266 [cs, stat]*.

Sherman, G. (2008). *Desktop GIS: Mapping the Planet with Open Source Tools*. Pragmatic Bookshelf.

Talbert, R. J. A. (2014). *Ancient Perspectives: Maps and Their Place in Mesopotamia, Egypt, Greece, and Rome*. University of Chicago Press.

Tallon, A. R. (2007). Bristol. *Cities*, 24(1):74–88.

Tennekes, M. (2018). Tmap: Thematic Maps in R. *Journal of Statistical Software, Articles*, 84(6):1–39.

The Economist (2016). The autonomous car's reality check. *The Economist*.

Thiele, J. (2014). R Marries NetLogo: Introduction to the RNetLogo Package. *Journal of Statistical Software*, 58(2):1–41.

Tobler, W. R. (1979). Smooth Pycnophylactic Interpolation for Geographical Regions. *Journal of the American Statistical Association*, 74(367):519–530.

Tomintz, M. N. M., Clarke, G. P., and Rigby, J. E. J. (2008). The geography of smoking in Leeds: Estimating individual smoking rates and the implications for the location of stop smoking services. *Area*, 40(3):341–353.

Tomlin, C. D. (1990). *Geographic Information Systems and Cartographic Modeling*. Prentice Hall, Englewood Cliffs, N.J.

Venables, W., Smith, D., and Team, R. C. (2017). *An Introduction to R. Notes on R: A Programming Environment for Data Analysis and Graphics*.

Venables, W. N. and Ripley, B. D. (2002). *Modern Applied Statistics with S*. Springer, New York, fourth edition.

Visvalingam, M. and Whyatt, J. D. (1993). Line generalisation by repeated elimination of points. *The Cartographic Journal*, 30(1):46–51.

von Wehrden, H., Hanspach, J., Bruelheide, H., and Wesche, K. (2009). Pluralism and diversity: Trends in the use and application of ordination methods 1990-2007. *Journal of Vegetation Science*, 20(4):695–705.

Wegmann, M., Leutner, B., and Dech, S., editors (2016). *Remote Sensing and GIS for Ecologists: Using Open Source Software*. Data in the Wild. Pelagic Publishing, Exeter. OCLC: 945979372.

Wickham, H. (2014a). *Advanced R*. CRC Press.

Wickham, H. (2014b). Tidy Data. *Journal of Statistical Software*, 59(10).

Wickham, H. (2016). *Ggplot2: Elegant Graphics for Data Analysis*. Springer, New York, NY, second edition.

Wieland, T. (2017). Market Area Analysis for Retail and Service Locations with MCI. *The R Journal*, 9(1):298–323.

Wilkinson, L. and Wills, G. (2005). *The Grammar of Graphics*. Springer Science+ Business Media.

Wise, S. (2001). *GIS Basics*. CRC Press.

Wulf, A. (2015). *The Invention of Nature: Alexander von Humboldt's New World*. Alfred A. Knopf, New York.

Xiao, N. (2016). *GIS Algorithms: Theory and Applications for Geographic Information Science & Technology*. London.

Zuur, A., Ieno, E. N., Walker, N., Saveliev, A. A., and Smith, G. M. (2009). *Mixed Effects Models and Extensions in Ecology with R*. Statistics for Biology and Health. Springer-Verlag, New York.

Zuur, A. F., Ieno, E. N., Saveliev, A. A., and Zuur, A. F. (2017). *Beginner's Guide to Spatial, Temporal and Spatial-Temporal Ecological Data Analysis with R-INLA*, volume 1. Highland Statistics Ltd, Newburgh, United Kingdom. OCLC: 993615802.

Index

Milton Keynes UK
Ingram Content Group UK Ltd.
UKHW051014071024
449327UK00012B/236